NATIONAL GEOGRAPHIC
SCIENCE

SCIENCE INQUIRY
AND WRITING BOOK

SCIENCE

PROGRAM AUTHORS

Randy Bell, Ph.D.

Malcolm B. Butler, Ph.D.

Kathy Cabe Trundle, Ph.D.

Judith S. Lederman, Ph.D.

David W. Moore, Ph.D.

NATIONAL GEOGRAPHIC LEARNING | CENGAGE Learning®

Science Inquiry

Life Science

Life Science

Science Inquiry

Life Science

Earth Science

Earth Science

Science Inquiry

Earth Science

Earth Science

Science Inquiry

Earth Science

Physical Science

Physical Science

Science Inquiry

Physical Science

Science in a Snap!

CHAPTER 1

Science in a Snap! Observe a Plant's Growth

Place a plant in a box that has a hole cut in one side. Close the box and put it in sunlight. After several days, **observe** the plant. Rotate the plant and close the box. **Predict** how the plant will grow. After several days, observe the plant again. How has the plant grown? **Infer** why it is growing that way.

Science in a Snap! Infer About Fossils

Observe the photos of the animal fossils. Use what you observe about their body parts and what you know about animals to answer each of these questions: Did the animal have a backbone? How did the animal move? Where did the animal live? What might the animal have eaten?

Science in a Snap! Observe Decomposers

You can use bread and water to grow decomposers. Sprinkle water on a piece of bread. Place the damp piece of bread in a resealable plastic bag. Seal and tape the bag shut. After several days, **observe** the bread with a hand lens. Do not open the bag. What is growing in the bag? How has the bread changed?

CHAPTER 4

Science in a Snap! Compare Animals

Observe the animals in the pictures. **Compare** their body parts. Find the following parts on one or more of the animals:

- Fur
- Wings
- Claws
- Scales

How do fur, wings, claws, and scales help animals survive?

When an animal hibernates, its heart beats less often. For example, when a woodchuck is not hibernating, its heart beats about 80 times in a minute. When the woodchuck hibernates, its heart beats very slowly. Use a stopwatch. Make a fist and squeeze your fingers together tightly one time every 15 seconds. That is how often a woodchuck's heart beats when it is hibernating. How many times does the hibernating woodchuck's heart beat in 1 minute? How might a slower heartbeat help a woodchuck survive when it hibernates?

Investigate Plants and Gravity

Question How does gravity affect the growth of plant roots?

Science Process Vocabulary

observe verb

When you **observe,** you use your senses to learn about an object or event.

predict verb

When you say what you think will happen, you **predict.**

I predict that the seeds will grow roots if I plant them in soil.

Materials

tape

2 plastic cups

8 paper towels

2 bean seeds

spoon

water

ruler

clay

What to Do

1 Use tape to label the cups **A** and **B.** Bunch up 4 paper towels in each cup. Place a bean seed in each cup right next to the side of the cup so you will be able to see its roots grow. Use a spoon to water the seeds. Add enough water to wet the whole paper towel.

2 Every other day, water the seeds. Add the same amount of water to each cup. **Observe** the seeds, watching for them to sprout. Record your observations in your science notebook.

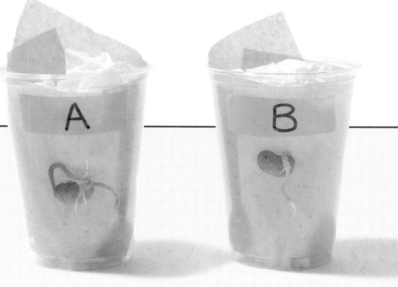

What to Do, continued

3 When the roots start to grow, wait until they are about 1.5 cm in length. Then carefully lay cup A on its side. Use the clay to hold cup A in place. Do not change the position of cup B. **Predict** in which direction the roots of each plant will grow. Record your predictions.

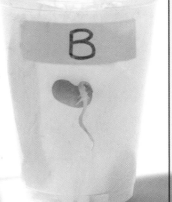

4 Continue to observe and water the plants every other day. When the roots have grown another 1.5 cm, carefully turn cup A upright. Do not change the position of cup B. Predict how the roots will grow. Record your prediction.

5 After 10 days, observe the roots in each cup. Record your observations.

Record

Write and draw in your science notebook.
Use a table like this one.

How Plant Roots Grow

Day	Cup A Observations	Cup B Observations

Explain and Conclude

1. Do the results support your **predictions?** Use your **observations** to explain.

2. **Infer** the way gravity affects the growth of plant roots. Tell what evidence from this activity you used to make your inference.

Parts of the roots of these cypress trees grow above water in this swamp.

Investigate Plant Parts

Question How can you classify plants by their parts?

Science Process Vocabulary

observe verb

You can **observe** objects or events by using one or more of your five senses.

classify verb

When you **classify,** you put things in groups according to their characteristics.

I can classify the plants by looking for things they have in common.

Materials

Picture Cards

hand lens

sunflower seed

corn seed

rye grass seed

microscope slide with spores

microscope

What to Do

1 **Observe** the sunflower, rye grass, and corn plants shown on the Picture Cards. Look for plant parts such as roots, stems, leaves, and flowers. Record your observations in your science notebook.

2 The sunflower, rye grass, and corn plants produce seeds. Use a hand lens and a microscope to observe the 3 different kinds of seeds. Record your observations.

Use these knobs to focus the microscope.

Place seeds here.

What to Do, continued

3 Observe the moss and the ferns on the Picture Cards. These plants do not produce seeds. They produce spores instead. Record your observations.

4 Use the microscope to observe the spores on the slide. Record your observations.

Place slide here.

5 Use your observations of the plants, seeds, and spores to **classify** the plants on Picture Cards 7–9 as either seed-producers or spore-producers.

Record

Write and draw in your science notebook.
Use tables like these.

Observations of Plant Picture Cards

Sunflower	
Rye grass	

Observations of Seeds and Spores

Corn seed	
Sunflower seed	

Classification of Plants

Plant	Seed-Producer or Spore-Producer?
Echinacea	

Explain and Conclude

1. How are the plants on all the Picture Cards similar and different?

2. How did your **observations** of seeds and spores help you to **classify** the plants on Picture Cards 7–9?

3. **Share** your classification with other groups. Talk about other ways that you could classify the plants on the Picture Cards.

Think of Another Question

What else would you like to find out about the parts of plants?

Investigate Animal Classification

Question How can you identify animals with backbones and model how a backbone works?

Science Process Vocabulary

classify verb

When you **classify**, you put things in groups according to their characteristics.

I can classify animals by observing how they are alike and different.

model noun

You can use a **model** to show how something in real life works.

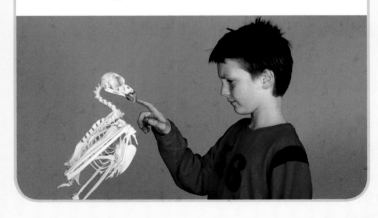

Materials

Animal Picture Cards

chenille stem

wooden beads

metal washers

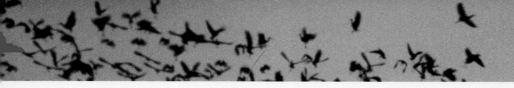

What to Do

1 **Observe** the animals on the Animal Picture Cards. Look at each animal's body parts, body coverings, and other physical features. Record your observations in your science notebook.

2 **Classify** the animals into 2 groups: animals that have backbones and animals that do not have backbones. Record your classifications.

What to Do, continued

3 You will make a **model** of a backbone.
Study the picture to see the parts of
a dog's backbone.

The backbone is made
of many small bones.
The spaces and discs in
between each bone allow
the backbone to move.

4 Choose one of the animals with a backbone from the
Picture Cards. Use the chenille stem, beads, and metal
washers to make a model of its backbone.

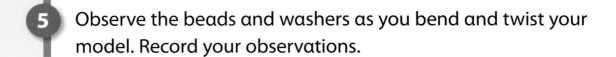

5 Observe the beads and washers as you bend and twist your
model. Record your observations.

Record

Write and draw in your science notebook.
Use a table like this one.

Animal Classification		
Animal	Observations	Backbone or No Backbone?
Chimpanzee		
Clownfish		

Explain and Conclude

1. How did you use your **observations** to **classify** the animals as animals with backbones and animals without backbones?

2. **Infer** how the structure of the backbone helps an animal move different ways. Use your observations of the **model** backbone to explain your answer.

Think of Another Question

What else would you like to find out about animals with backbones and how their backbones work? How could you find an answer to this new question?

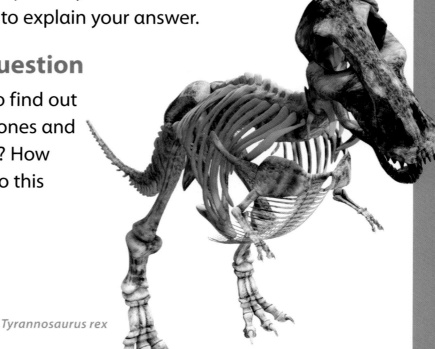

Tyrannosaurus rex

Investigate Fossils

Question How can you classify living things that lived long ago?

Science Process Vocabulary

Materials

observe verb

When you **observe,** you use your senses to learn about an object or event.

> I can feel that this fossil is rough.

fossils

classify verb

When you **classify,** you put things in groups according to their characteristics.

hand lens   ruler

Fossil Information Chart

paper

markers

What to Do

1 **Observe** all of the fossils. Choose one of them to study. Observe the fossil you choose with the hand lens. **Measure** its length. Examine its shape and other features. Record your observations in your science notebook.

2 Find your fossil on the Fossil Information Chart. Read about the animal that formed the fossil. Discuss with your group what you learned about the animal.

What to Do, continued

3 Make a card or poster about your fossil. First, draw a picture of your fossil. Include as many details as possible. Then, decide how you will add important facts about your fossil animal that you learned from the Fossil Information Chart. Write whether your fossil animal was a vertebrate or invertebrate. Tell about its environment.

Ammonite

4 **Share** your fossil card with the other groups. Discuss your drawing, your observations, and your **inferences** about the animal that formed the fossil.

Ammonite

This animal did not have legs.

5 After all groups have presented their fossil cards, think about how the fossil animals are alike and different. **Classify** the animals into vertebrates and invertebrates. Then choose another way to classify the fossils.

Record

Write and draw in your science notebook.
Use a table like this one.

Fossil	
Characteristic	Observations
Length	
Shape	

Explain and Conclude

1. Was your fossil animal an invertebrate or a vertebrate?
   What was its environment like? How do you know?

2. How else did you **classify** the fossils?

Think of Another Question

What else would you like to find out about classifying living
things that lived long ago? How could you find
an answer to this new question?

**British Columbia,
Canada**

This trilobite fossil was found in
Burgess Shale, a rock formation
that contains many fossils.

Investigate Interactions in a Model Pond

Question How do living things in a model pond ecosystem interact?

Science Process Vocabulary

model noun

You can make a **model** to show how something in real life works.

I can study a model pond to learn more about real ponds.

classify verb

When you **classify,** you put things in groups according to characteristics.

I can classify the plants as producers. The nutria is a consumer.

Materials

safety goggles

plastic bottle

sand

small rocks

Elodea plants

water

spoon

3 snails

hand lens

What to Do

1 Put on your safety goggles. Make a **model** of a pond ecosystem. Pour sand into the bottle so that the bottom is covered. Put rocks on top of the sand.

2 Place the *Elodea* plants in the bottle. Plant the bottom of the *Elodea* in the sand. Carefully pour water into the bottle until it is about two-thirds full. The top of the plant should float near the surface of the water.

3 Use the spoon to pick up 1 snail. Carefully place the snail in the water. Use the spoon to add 2 more snails to your model pond. Put your model in a sunny place.

What to Do, continued

4   **Observe** the parts of the model pond each day for 3 days. Decide whether each part of the model pond is living or nonliving. Use a hand lens to look closely at all parts of the model. Look for changes in both the living things and the nonliving things. Record your observations in your science notebook.

5   Use your observations to **infer** what each living thing needs and how it meets those needs. Does it produce its own food or consume energy by eating food? **Classify** each living thing as a producer or consumer. Write in your science notebook.

6   Draw your pond ecosystem. Circle the living things. Draw arrows to show how energy moves from the sun to the producers and consumers in the model pond ecosystem.

Record

Write and draw in your science notebook.
Use a table like this one.

Model Ecosystem

Model Pond Part	Observations	Living or Nonliving?	What I Infer About Its Needs	Producer or Consumer?
Sand				
Elodea plant				

Explain and Conclude

1. How did your **observations** help you **classify** producers and consumers in your ecosystem?

2. How did this **model** help you understand how living things in a real pond ecosystem interact?

My Model Pond Ecosystem

Think of Another Question

What else would you like to find out about how living things interact in a pond ecosystem? How could you find an answer to this new question?

A pond environment is home to **many producers and consumers.**

Investigate Brine Shrimp

Question How do different amounts of salt in the water affect hatching of brine shrimp eggs?

Science Process Vocabulary

variable noun

A **variable** is something that could change in an investigation. You change only one **variable** while you keep all the other parts the same.

I will only change how much salt I add to each cup.

estimate verb

When you **estimate,** you tell what you think about how much or how many.

There is about one scoop of brine shrimp eggs in the cup.

Materials

safety goggles

brine shrimp eggs

hand lens

3 cups of water

salt

measuring spoon

plastic spoon

craft stick with line

plastic wrap

36

Do an Experiment

Write your plan in your science notebook.

Make a Hypothesis

Brine shrimp are often found in saltwater environments. In this investigation, you will place brine shrimp eggs in water with different amounts of salt. How will the amount of salt affect the number of eggs that hatch? Write your **hypothesis.**

Identify, Manipulate, and Control Variables

Which variable will you change?
Which variable will you observe or measure?
Which variables will you keep the same?

What to Do

1. Put on your safety goggles. Use a hand lens to **observe** the brine shrimp eggs. Record your observations in your science notebook.

2. Label 1 cup with water **Control.** Use the measuring spoon to put 1½ teaspoons of salt in the cup. Stir gently with the plastic spoon until the salt is dissolved.

What to Do, continued

3 Choose how much salt you will add to the other 2 cups. You may add ½, 1, 2, 2½, or 3 measuring spoonfuls of salt. Label your cups to tell how much salt is in each cup. Use the measuring spoon to measure the salt and add it to the cups. Use the plastic spoon to stir gently until the salt is dissolved.

4 Dip the craft stick in the cup of brine shrimp eggs. Scoop out enough brine shrimp eggs to fill the end of the craft stick to the line. Add the eggs to the Control cup. Repeat for the other 2 cups. Cover each cup with plastic wrap.

5 Each day for 4 days, observe the cups. Look for newly hatched brine shrimp. **Estimate** the total number of brine shrimp that hatch in each cup. Record which cup has more hatched brine shrimp.

This brine shrimp just hatched.

Record

Write in your science notebook.
Use a table like this one.

Estimated Number of Brine Shrimp		
Cup 1: Control (1½ Spoonfuls of Salt)	Cup 2: ____ Spoonfuls of Salt	Cup 3: ____ Spoonfuls of Salt
Day 1		
Day 2		

Explain and Conclude

1. **Compare** the number of brine shrimp that hatched in each cup.

2. What can you **conclude** about how different amounts of salt affect the hatching of brine shrimp eggs?

Think of Another Question

What else would you like to find out about how different amounts of salt affect brine shrimp? How could you find an answer to this new question?

Cape Cod, Massachusetts

Brine shrimp live in saltwater environments like this salt marsh.

Investigate Plant Adaptations

Question How does a leaf's covering affect how quickly it wilts?

Science Process Vocabulary

predict verb

When you **predict**, you tell what you think will happen.

I predict that the plant will grow in sunlight.

infer verb

When you **infer**, you use what you know and what you observe to draw a conclusion.

I infer that the plant needs water.

Materials

leaf with waxy coating

leaf without waxy coating

paper towel

hand lens

What to Do

1 Place the 2 leaves on a paper towel. Use the hand lens to **observe** each leaf. Note each leaf's size, shape, color, and other properties. Record your observations in your science notebook.

2 Observe the surfaces of the leaves. Record your observations.

What to Do, continued

3 You will put both leaves in a sunny location. Use your observations to **predict** which leaf will wilt more quickly in sunlight. Record your prediction.

4 Place the leaves in a sunny location. Observe the leaves 3 times a day for 2 days. Record your observations.

Record

Write and draw in your science notebook.
Use a table like this one.

Comparing Leaves

Day and Time	Observations of Leaf With Waxy Coating	Observations of Leaf Without Waxy Coating
Start		

Explain and Conclude

1. Do the results support your **prediction?** Explain.

2. **Share** your results with others. Did they get the same results? Explain any differences.

3. **Infer** which kind of leaf covering is more likely to be found on plants that live in a dry place.

Think of Another Question

What else would you like to find out about how a plant's covering affects wilting? How could you find an answer to this new question?

Marsh plants near Okefenokee Swamp, Georgia

Math in Science

Graphing Data

Scientists use graphs to organize, summarize, and explain their data. The kind of graph a scientist uses depends on what kind of information he or she wants to show.

Line Plot A scientist observed the different kinds of birds in a field. She organized her data in a line plot. First, she made a column for each kind of bird she saw. Then, she placed an *X* in the correct column each time she saw that kind of bird.

Birds in Fletcher Field

Robin	Goldfinch	Blue jay	Red-winged blackbird
X			X
X			X
X	X		X
X	X	X	X
	X	X	X
	X	X	X

Pictograph A group of students started a paper recycling program at their school. They decided to collect data about how many bags of paper they recycled each week. They displayed their data in the pictograph below.

A pictograph is a way to compare information. Each picture, or symbol, stands for a certain amount. A key for the pictograph tells the amount each symbol stands for.

Bags of Paper Recycled

Week 1	🛍️ 🛍️ 🛍️ 🛍️ 🛍️
Week 2	🛍️ 🛍️ 🛍️ 🛍️ 🛍️ 🛍️
Week 3	🛍️ 🛍️ 🛍️ 🛍️
Week 4	🛍️ 🛍️ 🛍️ 🛍️ 🛍️ 🛍️ 🛍️

Key: 🛍️ = 2 bags

RECYCLE
PAPER

Line Graph Scientists use line graphs to show how something changes over time. The graph below shows how a kitten's mass changed over 8 weeks. The numbers along the left side of the graph show the kitten's mass in grams. The information along the bottom of the graph shows how old the kitten was. Each dot on the graph shows the kitten's mass at a certain age. The graph shows how the kitten's age and mass are related.

SUMMARIZE
What Did You Find Out?

1   Why do scientists use different kinds of graphs to organize data?

2   Which kind of graph would you use to show how the height of a seedling changes over a month? Why?

Graph Data

A scientist collected data about the number of seeds different sunflowers produced. The data are shown in the table.

Seeds and Flowers

Flower	Number of Seeds
1	100
2	400
3	800
4	500

1. Make a pictograph of the data. Start by writing the title at the top. Then make a row for each flower.

2. Decide what symbol you will use and how many seeds one symbol will stand for. Make a key.

3. To finish your graph, draw the correct number of symbols in each row.

Investigate Temperature and Coverings

Question How does a covering affect the temperature when a thermometer is placed in very cold water?

Science Process Vocabulary

hypothesis noun

You make a **hypothesis** when you state a possible answer to a question that can be tested by an experiment.

If there is no covering around the thermometer, then the temperature will not change.

infer verb

When you **infer,** you use what you know and what you observe to draw a conclusion.

I infer that the bison's fur keeps it warm.

Materials

2 thermometers

2 resealable plastic bags

cotton balls

feathers

fake fur

2 cups

very cold water

graduated cylinder

stop watch

Do an Experiment

Write your plan in your science notebook.

Make a Hypothesis

In this investigation, you will place two thermometers in cold water and measure the temperature. One of the thermometers will have a covering. You will choose the material for the covering: feathers, cotton balls, or fake fur. How will the covering affect the temperature of the thermometer? Write your **hypothesis.**

Identify, Manipulate, and Control Variables

Which variable will you change?
Which variable will you observe or measure?
Which variables will you keep the same?

What to Do

1 Place each thermometer in a plastic bag.

2 Choose the material you will use for a covering in your **experiment.** Place the material in the bag to cover 1 thermometer. Do not add anything to the other bag. Remove as much air as you can from the bags. Seal the bags.

What to Do, continued

3 Use the graduated cylinder to fill each cup with 100 mL of very cold water.

4 Place each bag in a cup of cold water. **Observe** the temperature of each thermometer. Record your **data** in your science notebook.

5 Predict what will happen to the temperature of each thermometer. Observe the temperature on the 2 thermometers every minute for 10 minutes. Record your data.

6 Make a graph of your data. Plot a line for the temperature of the thermometer with a covering. Plot a second line for the thermometer without a covering.

Record

Write and draw in your science notebook.
Use a table and a graph like the ones below.

Temperature Change in Thermometers		
Time	Temperature of Thermometer without Covering	Temperature of Thermometer Covered with _____
Start		
1 minute		

Explain and Conclude

1. **Compare** the temperatures of the two thermometers. Was your **hypothesis** supported? Explain.

2. **Share** your results with other groups. What type of covering kept the thermometers the warmest?

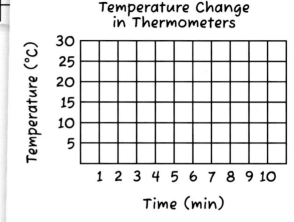

Temperature Change in Thermometers

3. Use the results of your **experiment** to **infer** how body coverings help animals live in cold temperatures.

Think of Another Question

What else would you like to find out about how coverings affect temperature? How could you find an answer to this new question?

The polar bear's thick fur and layer of fat help it survive in its icy environment.

Investigate Temperature and Cricket Behavior

Question How does temperature affect cricket behavior?

Science Process Vocabulary

data noun

Data are observations and information that you collect and record in an investigation.

share verb

When you **share** results, you tell or show what you have learned.

Materials

Stages in the Life of a Cricket chart

scissors

cricket habitat

thermometer

What to Do

1 Read the information on the Stages in the Life of a Cricket chart. Cut out the pictures and put them in order to show the cricket's life stages.

2 You can **observe** how temperature affects adult crickets by observing their behavior at different temperatures. **Measure** the temperature of the cricket habitat. Record the **data** in your science notebook.

3 Observe the activity of the crickets in their habitat. Pay special attention to how much they move. Record your observations.

4 Use the thermometer to measure the temperature in a refrigerator. Record your data.

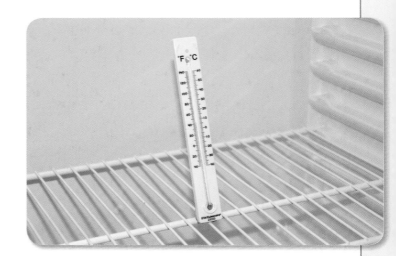

5 Place the cricket habitat in the refrigerator. **Predict** how the cooler temperature will affect the crickets' behavior. Record your predictions.

6 Remove the cricket habitat from the refrigerator after 10 minutes. Observe the cooled crickets in their habitat. Record your observations.

7 After 5 minutes record the temperature again. Observe and record the crickets' behavior.

Record

Write in your science notebook.
Use a table like this one.

Cricket Activity

Where Habitat Was	Temperature (°C)	Observations
Classroom		
Refrigerator (after 10 minutes)		

Explain and Conclude

1. **Share** your results with the class. Did your results support your **predictions?** Explain.

2. How did the change in temperature affect the crickets' behavior?

3. When the temperature of its environment gets cooler, the processes in a cricket's body slow down. **Infer** how this might affect a cricket's ability to move and survive.

Think of Another Question

What else would you like to find out about how temperature affects crickets' behavior? How could you find an answer to this new question?

This red cricket lives in wetland habitats.

Investigate Temperature and Seed Sprouting

Question **How does temperature affect seed sprouting?**

Science Process Vocabulary

variable noun

A **variable** is something that can change in an experiment.

In this experiment, I will change only one variable.

conclude verb

You **conclude** when you use information, or data, from an investigation to come up with a decision or answer.

Based on the results, I can conclude how temperature affects the growth of the seeds.

Materials

2 resealable plastic bags

tape

2 paper towels

spray bottle with water

seeds

2 thermometers

Do an Experiment

Write your plan in your science notebook.

Make a Hypothesis

In this investigation, you will grow seeds at different temperatures. How will temperature affect seed sprouting? Write your **hypothesis.**

Identify, Manipulate, and Control Variables

Which variable will you change?
Which variable will you observe or measure?
Which variables will you keep the same?

What to Do

1 Label the plastic bags **1** and **2.** Fold the paper towels and place one in each bag. Use the spray bottle to make the paper towels damp. Spray both paper towels the same number of times. Then put a thermometer in each bag.

2 Choose which type of seed you will test. Place 1 seed on the paper towel in each bag. Seal the bags.

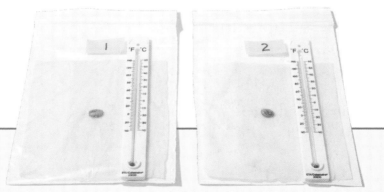

What to Do, continued

3 Place bag 1 on a flat surface in a dark place in the room. After 30 minutes, use the thermometer to **measure** the temperature. Record the **data** in your science notebook.

4 Decide whether you will put bag 2 in the refrigerator or the freezer. Place the bag on a flat surface in the place you choose. Wait 30 minutes and then measure the temperature of the bag. Record your data.

5 **Observe** both bags every other day for a week. Look for signs that your seeds are sprouting. Record your observations.

Record

Write and draw in your science notebook.
Use a table like this one.

Temperature and Seed Sprouting

Day	Observations	
	Bag 1: ____ °C	Bag 2: ____ °C
Start: After 30 minutes		
1		
3		

Explain and Conclude

1. Which seed had grown more by the end of the week?

2. What can you **conclude** about temperature and how seeds sprout?

3. How might seed growth be affected in places that have cold winters?

Think of Another Question

What else would you like to find out about temperature and seed sprouting?

Snake River and the Teton Range, Grand Teton National Park, Wyoming

Do Your Own Investigation

**Question** **Choose one of these questions, or make up one of your own to do your investigation.**

- How do light and darkness affect how seeds sprout and grow?
- How does temperature affect how long it takes for a butterfly pupa to change to an adult?
- What will happen if you move a cactus to a wetland environment?
- Are pill bugs attracted to a light environment or a dark environment?
- How does increasing the amount of light affect the growth of a plant?

Science Process Vocabulary

hypothesis noun

You make a **hypothesis** when you state a possible answer to a question that can be tested by an experiment.

I will test the hypothesis that cold temperatures slow down the growth of an insect.

Open Inquiry Checklist

Here is a checklist you can use when you investigate.

- ☐ Choose a **question** or make up one of your own.

- ☐ Gather the materials you will use.

- ☐ If needed, make a **hypothesis** or a **prediction.**

- ☐ If needed, identify, manipulate, and control **variables.**

- ☐ Make a **plan** for your **investigation.**

- ☐ Carry out your **plan.**

- ☐ Collect and record **data. Analyze** your data.

- ☐ Explain and **share** your results.

- ☐ Tell what you **conclude.**

- ☐ Think of another question.

Four Peaks Mountain, Sonoran Desert, Arizona

Desert plants have adaptations that help them to survive in a very dry environment.

Write
Like a Scientist

Write About an Investigation

Surviving in a Changing Environment

The following pages show how one student, Tiara, wrote about an investigation. As she read about living things and their environments, Tiara became interested in finding out whether a desert plant could survive in a wetland environment. Here is what she thought about to get started:

- Tiara wanted to work with living things that she could observe in the classroom, so she decided to use cacti in her investigation.
- She needed to use materials that she could obtain easily and work with safely. She would base her question on the materials she could use and the environment she would create.
- Tiara decided to build a model wetland habitat and place a cactus in it.
- She would observe the cactus to see whether it could grow in its new habitat.

Desert

A desert environment is very dry.

Wetland

A wetland environment is very wet.

Model

Question

What will happen if you move a cactus to a wetland environment?

> Make sure the question asks exactly what you are trying to find out.

Materials

2 cactus plants (same kind, same size)

hand lens

large resealable plastic bag

spray bottle with water

> Carefully describe the objects and materials you will need to answer your question. Be specific about the materials you will use.

Your Investigation

Now it's turn to do your investigation and write about it. Write about the following checklist items in your science notebook.

☐ Choose a question or make up one of your own.

☐ Gather the materials you will use.

Model

My Hypothesis

If I place the cactus in a wetland environment, then it will begin to turn yellow and die. A similar cactus that is not in a wet environment will stay green and healthy.

State what you think will happen in your experiment. Using an "If..., then..." statement is a good way to make your hypothesis clear.

Your Investigation

☐ **If needed, make a hypothesis or prediction.**

Write your hypothesis or prediction in your science notebook.

Model

Variable I Will Change

One cactus will be in a dry environment. The other will be in a damp environment.

Variable I Will Observe

I will observe to see if either cactus starts to change color or look unhealthy.

> During an investigation, it is important to accurately record observations.

Variables I Will Keep the Same

Everything else will be kept the same. Both plants will be the same kind and size. They will be in the same kind of container and have the same kind and amount of soil. Both will be placed in the same location.

Your Investigation

☐ If needed, identify, manipulate, and control variables.

Write about the variables in your investigation.

Model

My Plan

1. Label the cacti **Desert** and **Wetland**.
2. Observe the 2 cacti with a hand lens.
3. Spray the Wetland cactus with water. Spray until the water makes the soil very moist. Spray the Desert cactus one time.
4. Place the Wetland cactus in a plastic bag and seal the bag.
5. Place both plants in a sunny place.
6. Every day for 3 weeks, spray the Wetland cactus 3 times with the water. Spray the Desert cactus only on the first day. Record your observations each day.

> Your plan should tell everything you will do. Give exact times or amounts when necessary.

Your Investigation

☐ **Make a plan for your investigation.**

Write the steps for your plan.

Model

I carried out all six steps of my plan.

 **Your Investigation**

☐ **Carry out your plan.**

Be sure to follow your plan carefully.

You might make a note if you needed to adjust your plan in any way. Tiara found she did not need to adjust her plan.

Model

Data (My Observations)

Day	Observations	
	Desert Cactus	Wetland Cactus
1	Cactus is green with light colored spines	Cactus is green with light colored spines. Environment in bag is damp.
21	no change	turning yellow

> Use a table to organize your observations.

My Analysis

At the start of the experiment, both cacti looked about the same. After 3 weeks, the cactus in the wetland environment began to turn yellow.

> Explain what you observed during the entire investigation.

Your Investigation

☐ **Collect and record data. Analyze your data.**

Collect and record your data, and then write your analysis.

Model

How I Shared My Results

I took photos of the cacti every day. I used the photos to make a computer presentation. I shared the presentation with classmates.

> Using pictures, diagrams, and graphs can help others understand your results.

My Conclusion

The cactus in the desert environment did not change. The cactus in the wetland environment turned yellow and started to die. The results of the investigation support my hypothesis.

> Make a conclusion based on your data and observations. Check whether your results support your hypothesis.

Another Question

I wonder what would happen to a wetland plant if it were placed in a desert environment. Would it die? Or would it just not grow as well as in its wetland environment?

> One investigation often gives you ideas for other investigations.

Your Investigation

☐ Explain and share your results.

☐ Tell what you conclude.

☐ Think of another question.

A plant system

How
Scientists Work

Using Models to Study Systems

A system is a group of parts that work together. Some systems that scientists might study include the parts of a plant, the human digestive system, or a river system in the United States.

Sometimes scientists use models to study how systems work. Models might look like the real thing, or they might be very different.

Scientists might use models to study different parts in an ecosystem. An ecosystem includes all the living and nonliving things in a specific area.

Computer model of a river system

A pond is an ecosystem.

Diagram One kind of model scientists use to study ecosystems is a diagram. A food chain diagram shows how one kind of organism gets energy from another kind of organism. A food chain diagram for a pond ecosystem might look like this:

sun → plant → fish → frog → hawk

Map A map is another kind of model that scientists might use to study a pond. The map could help scientists figure out how the water level changes in a pond.

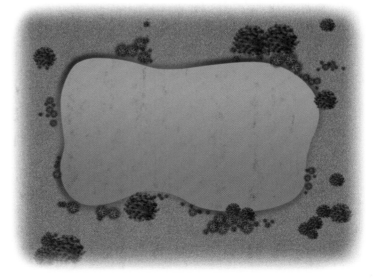

Map of a pond

Physical Models Scientists might also build physical models of the pond. They could use a container, rocks, plants, and animals to make a model ecosystem. It might look like the one in the picture.

Scientists could use this model to find out what would happen to the plants in a real pond if they added more fish. Here is what they might do:

1 Build a model pond with the same kinds of fish that are in the real pond.

2 Observe the model for one week to see what happens to the plants and fish.

3 Add more fish.

4 Observe the plants and fish in the model for one week.

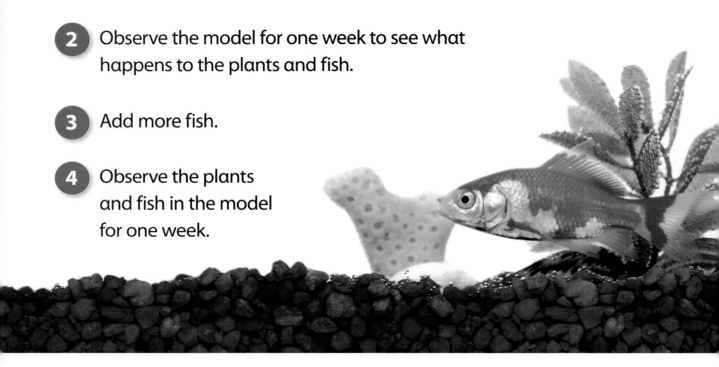

SUMMARIZE
What Did You Find Out?

1 What is a system?

2 What are three kinds of models scientists might use to study an ecosystem?

Plan a Model

The map below shows a pond near family homes. Many of the people who own the homes use salt on their sidewalks in winter. When rainwater runs over the land, it carries some of the salt to the pond.

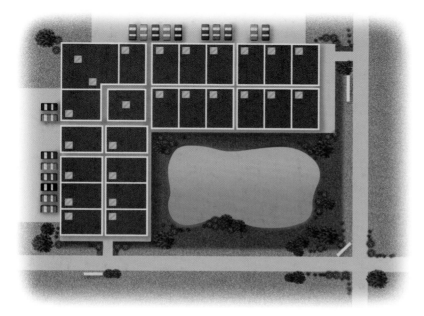

How could you use a model to study how the salt water might affect the pond?

1. Identify the question you are trying to answer.

2. Which kind of model or models would you use?

3. Write a procedure for using the model to answer the question.

4. Share your plan with others. Talk about changes you might make to your plan.

Science in a Snap!

CHAPTER 1

Science in a Snap! Make a Moonlight Model

Place a globe on a table. The globe is a **model** of Earth. Have your partner stand behind the table and hold a foam ball behind and slightly above the globe. The foam ball is a model of the moon. Turn off the lights and turn on a flashlight to light the globe and ball. The flashlight represents the sun. How much of the model moon and model Earth are lighted by the model sun? Of the sun, moon, and Earth, which produce light and which reflect light?

Science in a Snap! **Compare Sunlight and Shade**

Stand in the sunlight for 3 minutes. Tell how you feel and what you see. Move to the shade. Tell how you feel and what you see. **Compare** how you feel in the sunlight to how you feel in the shade. What causes the differences you feel and see?

Science in a Snap! **Observe Differences in Stars**

Look at the picture of the stars. All stars except the sun look like points of light because they are so far away. Try to find some bright stars. Now look for faint, or dim, stars. Choose a section of the picture and **count** the number of different colored stars you can see. Explain how a large star could appear dimmer than a smaller star. Explain how two stars that are alike can appear to have different brightnesses.

CHAPTER 4

Science in a Snap! Observe the Minerals in a Rock

Use a hand lens to **observe** a piece of granite rock. Look for mineral particles of different colors, shapes, or sizes. How many different minerals do you think are in your rock? **Compare** your observations with a partner. How are the rocks alike and different?

CHAPTER 5

Science in a Snap! List Uses for Natural Resources

Metals are important natural resources. Make a list of all the ways you can think of that people use metals. Then **compare** your list with a partner's list. What is on your list that is not on your partner's list? Discuss with your partner whether metals are renewable or nonrenewable resources.

Science in a Snap! Erosion in Action

Make a tube out of foil. Put some sand along the center of the tube and pat it down. Place one end of the tube over the cup. Hold the other end of the tube a little higher. You have made a **model** of a river. Slowly pour 50 mL of water down the tube so that the water runs into the empty cup. What happens to the sand? **Observe** the water in the cup. How does this model show how water can move materials in a river?

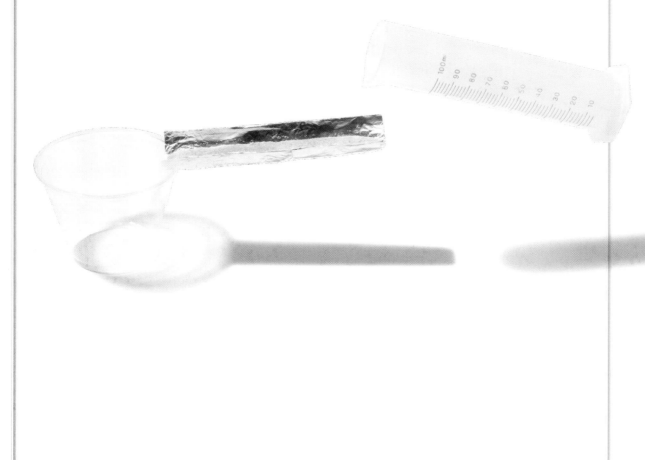

Science in a Snap! Model Underwater Earthquakes

When an earthquake occurs under the ocean, it can cause a very tall, powerful wave called a tsunami. Make a **model** to show how a tsunami happens. Fill a plastic bin halfway with water. Hold 2 wooden blocks together under the water in the center of the pan. Quickly move them against each other in opposite directions and **observe** what happens to the water. How might a tsunami affect the land that is near the ocean?

Science in a Snap! Conserving Water

The container in the photo holds 1 gallon of water. The average person in the United States uses about 75 gallons of water each day. Work with a partner. Use the table to help you make a list of ways that you can use less water each day. Think of other ways, too. **Compare** your list with other groups. How much water could your class save each week?

Daily Water Use in a House of 4 People	Amount of Water Per Day (gallons)
Showering/ brushing teeth	44
Flushing toilets	25
Doing laundry	14
Washing dishes	4

Investigate Moon Phases

Question How can you model the way the moon's shape seems to change?

Science Process Vocabulary

model noun

A **model** can show how things in real life work.

analyze verb

You **analyze** data when you look for patterns and relationships in the data.

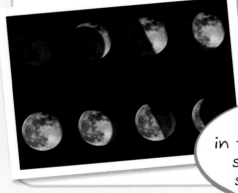

I see a pattern in the way the moon seems to change shape over time.

Materials

Calendar of Moon Phases

safety goggles   scissors

Moon Phases Cut-Outs

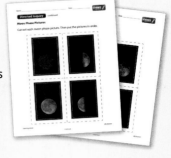

glue   index cards

clip

What to Do

1 **Observe** the Calendar of Moon Phases. Is there any pattern to the way the moon looks? Describe the pattern in your science notebook.

2 Put on your safety goggles. Cut out the pictures from the Moon Phases Cut-Outs.

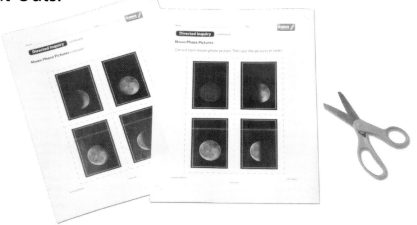

3 Glue each moon picture to an index card. Be sure to place the picture at the bottom of the card as shown in the photo.

What to Do, continued

4 Think of the patterns you observed in the moon phases in step 1. Put the moon pictures in order according to that pattern.

5 Clip the cards together at the top to make a moon phase flipbook. The flipbook is a **model** of how the moon's shape seems to change.

6 With one hand, hold the cards at the top near the clip. With the other hand, quickly flip through the cards from front to back. Observe the moon pictures as you flip through the cards. How does the shape of the lighted part of the moon appear to change? Draw your observations in your science notebook.

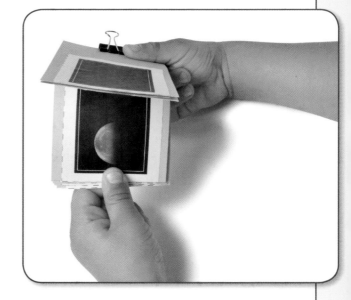

Record

Draw how the moon's shape
appears to change.

Changes in the Moon's Shape in One Month

Explain and Conclude

1. What pattern do you see in your **model** of moon phases?

2. Describe how your model shows the phases of the moon.

3. Suppose you are able to see the whole lighted part of the moon
   in the night sky. Use your model to **predict** how the shape of the
   lighted part of the moon will change over the next week.

Mono Lake,
California

Math in Science

Bar Graphs

Scientists use a bar graph to show data that are not changing over time. A bar graph has a title that tells what the graph is about. You can tell by the title that the graph below is about the distance different model rockets traveled.

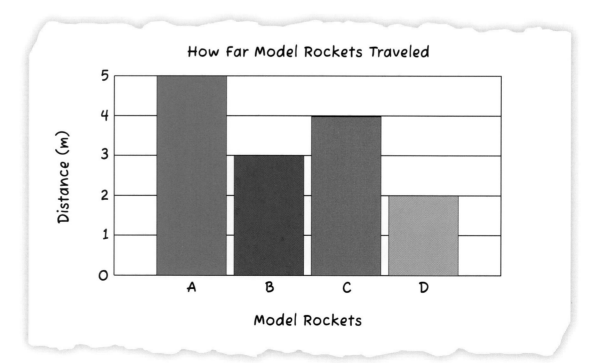

How Far Model Rockets Traveled

The labels across the bottom of the graph identify the different rockets as A, B, C, or D. A colored bar stands for each rocket.

The rocket that traveled the farthest has the tallest bar. The one that traveled the least distance has the shortest bar.

You can also tell exactly how far each rocket traveled. Look at the numbers along the left side of the graph. This scale tells how many meters were traveled. You know that the numbers stand for meters because the **(m)** at the end of the label **Distance** tells you that. You can see that Rocket A traveled 5 m. Rocket D only traveled 2 m.

Making a Bar Graph

You can use a bar graph to organize information you collect during science investigations. Follow these steps.

1 Decide what data you will show. You might want to show how far, how many, or how much of something.

2 Write a title for your graph.

3 Write numbers on the side of the graph, starting at the bottom. Write a label to tell what the numbers mean.

4 Label the bottom of the graph to tell what the different bars stand for.

5 Draw and label your bars. Make each bar a different color.

SUMMARIZE
What Did You Find Out?

1 What kind of data would scientists show on a bar graph?

2 How can you tell what the different bars on a graph stand for?

Make and Use a Graph

Look at one student's data below about the rocks he collected in a particular area. Then use the data to make a bar graph.

Kind of Rock	How Many Collected
Granite	1
Shale	3
Sandstone	5
Pumice	2

1 Write a title for your graph.

2 Write numbers on the side of the graph and write **How Many Collected.**

3 Write **Kind of Rock** on the bottom of the graph.

4 Write the name of each rock along the bottom. Draw a bar to show how many of each rock were collected.

Share your graph with a partner. Ask and answer questions about your graphs.

Investigate Sunlight and Shadow

Question How does a shadow caused by sunlight change during the day?

Science Process Vocabulary

observe verb

You **observe** when you use your senses to learn about an object or event.

I feel cooler in the shade of the tree.

predict verb

You **predict** when you tell what you think will happen.

I predict the shadow will change during the day.

Materials

white paper

tape

toy figure

ruler

colored pencil

What to Do

1 Tape the paper in a sunny spot. Put the toy figure near the top of the paper. **Observe** the shadow the toy makes.

2 Mark an **X** on the paper that shows the position of the sun. Record how high the sun looks in the sky. Write the date and time next to the X. Then trace the toy's shadow. Write the date and time next to it. Also record the date and time in your science notebook.

3 Use the ruler to **measure** the length of the shadow. Record your measurement in your science notebook.

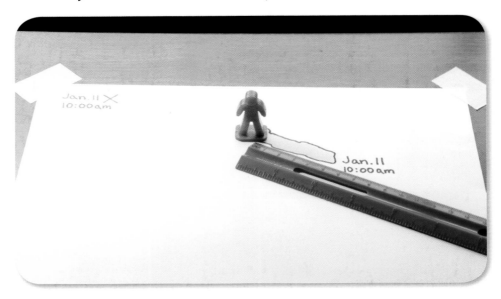

4 Repeat steps 2 and 3 at three more times during the day. Observe how the shadows change throughout the day.

5 Look at your **data. Predict** the shadow's position 1 hour after your last observation. Use the colored pencil to draw where you think it will be. Mark an **X** in colored pencil where you think the sun might appear to be. After 1 hour, observe the shadow to check your prediction.

Record

Write or draw in your science notebook.
Use a table like this one.

Changes in Shadows

Date	Time	Sun's Position in the Sky	Length of Shadow (cm)

Explain and Conclude

1. **Compare** your measurements. When was the shadow the shortest and longest? Where was the sun in the sky at those times?

2. What pattern in length and movement did you **observe** with the shadows?

3. What caused the changes you observed in the shadows?

Think of Another Question

What else would you like to find out about sunlight and shadows? How could you find an answer to this new question?

Saint Michael's Mount, Cornwall, England, United Kingdom

You can look at the shadow on this sundial to tell what time it is.

Investigate Energy from the Sun

Question What happens to the temperature of water when it is in sunlight and in the shade?

Science Process Vocabulary

measure verb

When you **measure,** you find out how much or how many.

I can use a thermometer to measure temperature.

compare verb

When you **compare,** you tell how objects or events are alike and different.

The temperature of the hot tea is higher than the temperature of the cold juice.

Materials

2 plastic cups   tape

water   graduated cylinder

2 thermometers   stopwatch

What to Do

1 Label 2 cups **Sun** and **Shade.**

2 Use the graduated cylinder to **measure** 75 mL of water. Carefully pour the water into the Sun cup. Repeat with the Shade cup.

3 Place a thermometer in each cup. Use a stopwatch to time 1 minute. Then measure the temperature of the water in the cups. Record your **data** in your science notebook.

4 Place the Sun cup in bright sunlight. Place the Shade cup in the shade. Wait 1 hour. Measure the temperature of the water in each cup again. Record your data.

5 Move the Sun cup to the shady place. Wait 1 hour. Then measure the temperature of the water in both cups. Record your data.

Record

Write in your science notebook.
Use a table like this one.

Water Temperature			
Cup	At Start (°C)	After 1 Hour (°C)	After 2 Hours (°C)
Sun		In sunlight	In shade
Shade		In shade	In shade

Explain and Conclude

1. **Compare** the temperatures of the water in the 2 cups at the end of 1 hour in step 4.

2. What happened to the temperature of water in the Sun cup when it was moved from the sunlight to the shade? Explain why you think this happened.

3. **Share** your **data** with others. Explain any differences in the data.

Think of Another Question

What else would you like to find out about sunlight and water temperature? How could you find an answer to this new question?

Will the iced tea be warmer or colder after sitting in sunlight?

Investigate Sunlight

Question How well do different materials block sunlight?

Science Process Vocabulary

variable noun

A **variable** is a part of an experiment that you can change.

In an experiment, you change only one variable while you keep all the other variables the same.

> I will change only one material in my investigation.

Materials

light-sensitive beads

stopwatch

foil

wool cloth

paper towel

plastic bag

bag with sunscreen

paper

Do an Experiment

Write your plan in your science notebook.

Make a Hypothesis

In this experiment, you will investigate rays from the sun. You will use light-sensitive beads, which change color in sunlight. Then you will choose two materials to place in front of the light-sensitive beads. Which materials block sunlight? Write your **hypothesis.**

Identify, Manipulate, and Control Variables

Which variable will you change?
Which variable will you observe?
Which variables will you keep the same?

What to Do

1 Place 10 light-sensitive beads in a shady place. **Observe** the beads. Record your observations in your science notebook.

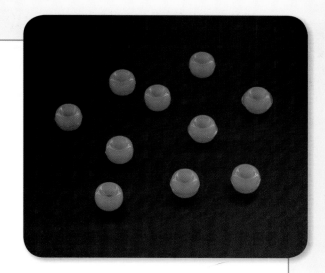

2 Put 5 of the beads in a sunny place. Wait 2 minutes, and observe what happens to the beads. Record your observations. Remove the beads from the sunlight.

3 Choose 2 materials to place between the beads and the sunlight. Record your choices.

4 Hold the 2 materials in front of a window at the same time. Have a partner place 5 beads behind each material. Wait 2 minutes. Then observe the beads. Record your observations.

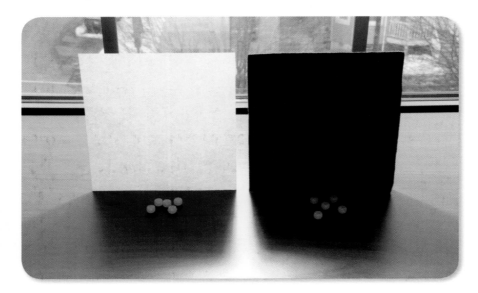

5 **Compare** your results with groups that tested different materials. Which materials blocked the most sunlight? Which materials blocked the least sunlight?

Record

Write in your science notebook.
Use a table like this one.

Light-Sensitive Beads

Location	Material Tested	Observations of Beads (no change, slight change, full change)
Shade		
Sunlight		

Explain and Conclude

1. Do your results support your **hypothesis?** Explain.

2. **Compare** what happened to the beads in steps 1 and 2. What caused the change in step 2?

3. What can you **conclude** about how well materials block sunlight? What evidence did you use to come to your conclusion?

Think of Another Question

What else would you like to find out about how to block sunlight? How could you find an answer to this new question?

Sesma Navarra, Spain

These solar panels absorb energy from sunlight.

Investigate Light Brightness

Question How does a light's brightness appear to change with distance?

Science Process Vocabulary

predict verb

When you **predict,** you tell what you think will happen.

I predict that the flashlight will not look as bright if I turn on the lights.

conclude verb

You **conclude** when you use information or data from an investigation to come up with a decision or answer.

I can use my results to conclude which light appears brightest.

Materials

3 penlights

meterstick

tissue paper

What to Do

1 Label the penlights **A, B,** and **C.** Penlights A, B, and C represent stars that are similar in size and temperature to the sun. Have 3 partners **measure** 2 m away from you. Have them stand at that distance, each holding a light. Have each partner cover the light with tissue paper. The lights should be pointing toward you but should not be turned on.

2 **Predict** how bright each light will look when your partners turn on their penlights. Record your predictions in your science notebook. Have your partners turn on the lights. **Observe** how bright they appear. Record your observations. Use the Apparent Brightness Scale to describe brightness.

Apparent Brightness Scale	
1	very bright
2	bright
3	dim

3 For trial 1, have your partners measure and stand at the following distances. Penlight A should be 2 m away from you. Penlight B should be 4 m away from you. Penlight C should be 3 m away. Then repeat step 2.

4 For trial 2, have your partners measure and stand at the following distances. Penlight A should be 2 m away from you. Penlight B should be 3 m away from you. Penlight C should be 4 m away. Then repeat step 2.

5 For trial 3, change the positions of the lights one more time. Penlight A should be 2 m away. Penlight B should be 5 m away. Penlight C should be 4 m away. Then repeat step 2.

Record

Write in your science notebook.
Use a table like this one.

Distance and Apparent Brightness of Lights

	Penlight	Distance from Observer	Predicted Brightness	Observed Brightness
Start	A	2 m		
	B	2 m		
	C	2 m		
Trial 1	A	2 m		
	B	4 m		
	C	3 m		

Explain and Conclude

1. Do your results support your **predictions?** Explain.

2. What can you **conclude** about distance and the apparent brightness of lights that are the same size? Use your **observations** to support your conclusion.

3. Use the results of this investigation to explain how stars that are like each other in size and temperature can appear to have different brightnesses.

Think of Another Question

What else would you like to find out about how a light's brightness appears to change with distance? How could you find an answer to this new question?

Investigate Lenses

 **Question** How can lenses help you see objects that are far away?

Science Process Vocabulary

observe verb

When you **observe,** you use your senses to learn about an object or event.

compare verb

When you **compare,** you tell how objects or events are alike and different.

One telescope is older. Both telescopes make it easier to see more stars.

Materials

3 lenses

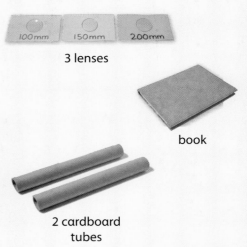
book

2 cardboard tubes

tape

Star Photo

meterstick

What to Do

1 **Observe** the words in the book with all 3 lenses. **Compare** how large the words look with each lens. Record your observations in your science notebook.

2 Choose 2 of the lenses to make a telescope, and record your choice. Hold 1 lens carefully in the center on the end of the tube and tape the lens in place. Be careful that the lens does not fall through the tube. Repeat with the other lens and tube. Then slide the tubes together. Make sure the lenses are facing out.

What to Do, continued

3 Tape the Star Photo to a wall. **Measure** 3 m from the photo. Stand at that place. Observe the stars in the picture without the telescope. Record your observations.

4 Use your telescope to observe the stars. Slide the cardboard tubes in and out until you see a clear image. Record your observations.

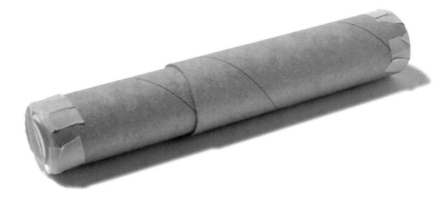

5 Turn the tube around so that the lenses are reversed. Observe the stars. Record your observations.

6 Exchange your telescope with another group. Repeat step 4. Compare what you see through the second telescope with what you saw through your telescope.

Record

Write in your science notebook.
Use a table like this one.

Observing Stars

	Observations
Without telescope	
With telescope	
With telescope reversed	

Explain and Conclude

1. How did turning the telescope around affect the stars you **observed?**

2. **Compare** the number of stars you could observe with your telescope and without a telescope.

3. Use your observations to explain how lenses can help scientists study stars.

Think of Another Question

What else would you like to find out about how lenses can help you see objects that are far away? How could you find an answer to this new question?

Scientists use telescopes to study objects in space, such as Comet Hyakutake.

Investigate Rock Layers

Question How can you model and compare rock layers?

Science Process Vocabulary

model noun

You can make and use a **model** to show how something in real life works.

compare verb

You **compare** when you tell how things are alike and different.

> I can compare rock layers.

Materials

safety goggles

safety mask

plaster of Paris

measuring spoon

3 bowls

graduated cylinder

water

3 plastic spoons

sand

pebbles

food coloring

paper cup

stopwatch

sandpaper

hand lens

What to Do

1 Put on your safety goggles and mask. Use the measuring spoon to put 3 spoonfuls of plaster of Paris into a bowl. **Measure** 20 mL of water with the graduated cylinder. Use the plastic spoon to slowly stir the water into the plaster of Paris.

2 Add the sand and 5 drops of green food coloring to the plaster of Paris mixture. Stir until the mixture is blended.

3 Use the plastic spoon to put the mixture into the paper cup. This is the bottom layer of a **model** rock. Write in your science notebook. Tell what materials you used to make this model rock layer.

4 Repeat steps 1 and 2 to make a middle rock layer. This time, use plaster of Paris, pebbles, red food coloring, and a new bowl and spoon. Pour the mixture over the bottom layer.

5 Repeat steps 1 and 2 again to make a top layer. This time, use only plaster of Paris, water, and blue food coloring, as well as a new bowl and spoon. Pour the mixture over the bottom layers.

6 Wait 30 minutes for the rock model to set. Then tear away the paper cup to remove the model. Place the sandpaper on your desk. Rub the side of the model over the sandpaper 5 times. **Observe** the layers with the hand lens. Feel the layers. Record your observations.

Record

Write or draw in your science notebook.
Use a table like this one.

Rock Model

Layer	Materials Used	Observations
Bottom		
Middle		
Top		

Explain and Conclude

1. Describe the layers of your **model.**

2. **Compare** your model rock to those of other students. Why do you think the models look different?

3. How is your model rock like the real rocks in the picture on this page? How is it different?

Think of Another Question

What else would you like to find out about rock layers? How could you find an answer to this new question?

Colorado Plateau

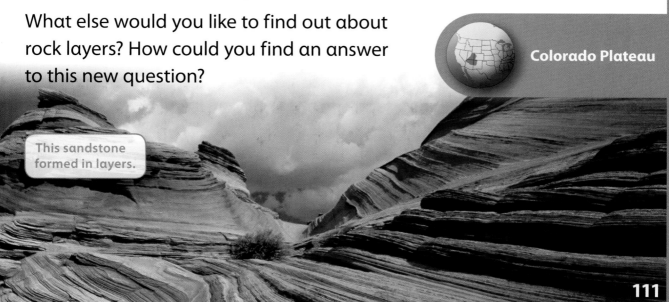

This sandstone formed in layers.

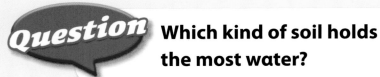

Guided Inquiry

Investigate Soil and Water

Question Which kind of soil holds the most water?

Science Process Vocabulary

variable noun

A **variable** is a part of an experiment that can change.

You change only one **variable** while you keep all the other parts the same. You control the parts that do not change.

The only variable I will change is the kind of soil in the cup.

Materials

sandy soil · clay soil

peat · hand lens

measuring cup · 2 cups with holes

pan balance · gram masses

2 cups without holes · graduated cylinder

water · stopwatch

112

Do an Experiment

Write your plan in your science notebook.

Make a Hypothesis

In this experiment, you will investigate the amount of water that different types of soils can hold. You will choose two kinds of soil to test: sandy soil, clay soil, or peat. Then you will compare your results with those from all groups. Which kind of soil will hold the most water? Write your **hypothesis.**

Identify, Manipulate, and Control Variables

Which variable will you change?
Which variable will you observe?
Which variables will you keep the same?

What to Do

1 Choose 2 kinds of soil. **Observe** each kind of soil with the hand lens. Record your observations.

2 Use the measuring cup to **measure** and pour 75 mL of each kind of soil into its own plastic cup with holes.

What to Do, continued

3 Use the pan balance and gram masses to find the mass of each cup of soil. Record your **data.**

4 Place each cup of soil inside a cup without holes. Then use the graduated cylinder to slowly pour 50 mL of water in each cup of soil.

5 Wait 10 minutes. Then use the balance to measure the mass of each cup of wet soil. Record your data.

6 Subtract the mass of the cup with dry soil from the mass of the cup with wet soil to find out how much water each soil held.

Record

Write in your science notebook.
Use a table like this one.

Soil and Water

	Kind of Soil _____	Kind of Soil _____
Observations		
Mass of dry soil and cup (g)		
Mass of wet soil and cup (g)		
Mass of water in soil (g)		

Explain and Conclude

1. Do your results support your **hypothesis?** Explain.

2. **Compare** your results with other groups. Which kind of soil holds the most water?

3. **Infer** why soils that hold a medium amount of water would be good for growing plants.

Think of Another Question

What else would you like to find out about soil and water?
How could you find an answer to this new question?

Investigate Natural Resources

Question How can you identify and classify natural resources?

Science Process Vocabulary

classify verb

You **classify** when you put things in groups by how they are alike and different.

I can classify objects by the type of natural resources they are made from.

compare verb

You **compare** when you tell how things are alike and different.

These materials are all made of plastic, but they have different shapes.

Materials

pencil

metal washer

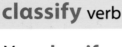
eraser

rock

soil

water

air

aluminum foil

spoon

index cards

markers

2 sheets paper

What to Do

1 **Observe** the items. List them and some of their uses in your science notebook.

2 Use the red marker to write the name of each object at the top of an index card.

Eraser

3 What natural resource was used to make each object? Use the black marker to write the name of the natural resource on each item's card. If the item was made from more than one resource, make a separate card for each natural resource.

Eraser
Rubber.

What to Do, continued

4 Label 1 sheet of paper **Nonrenewable.** Label the other sheet **Renewable.**

5 **Classify** the resource on each card. Put the cards with resources that can be used up on the Nonrenewable paper. Place the cards with resources that cannot be used up on the Renewable paper. Record your classifications in your science notebook.

Nonrenewable

Renewable

Eraser
Rubber

6 Count the number of cards in each group. Use your **data** to make a graph that shows the number of renewable and nonrenewable resources you found.

Record

Write in your science notebook.
Use a table like this one.

Resource Uses and Classification

Resource	Uses	Renewable or Nonrenewable?
Rubber		

Explain and Conclude

1. Which resource has the most uses? Is this resource renewable or nonrenewable?

2. **Compare** your **data** table with the class. Did you **classify** resources the same? What might you add to your table?

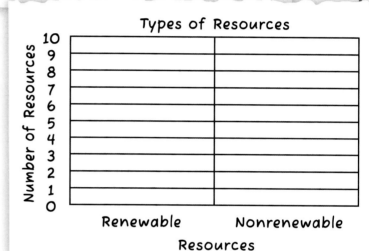

3. Look at the nonrenewable resources you found. How might you use fewer of these resources every day?

Think of Another Question

What else would you like to find out about natural resources? How could you find an answer to this new question?

What natural resources do you see in this picture?

Investigate Wind Energy

Question How can you use wind energy to move objects?

Science Process Vocabulary

model noun

A **model** can show how something in real life works. A model may or may not look like the real thing.

compare verb

When you **compare**, you tell how things are alike and different.

Materials

safety goggles

5 spoon pieces

2 lumps of clay

straw

index card

tape

string

cup with hole

paper clip

pennies

What to Do

1 Put on your safety goggles. Choose whether you will use 3, 4, or 5 spoons to make a windmill. Record your choice in your science notebook. Stick the spoons into the larger lump of clay. Push the straw into the clay. You have made a **model** windmill.

Make sure all the spoons face the same direction and that they are evenly spaced.

2 Use the tape and the index card to attach your windmill to a desk or table as shown in the picture.

Tape the card down lightly to make sure that the windmill can still turn.

What to Do, continued

3 Divide the small piece of clay into 4 pieces. Put these pieces of clay on both sides of the straw. The clay will keep the straw from sliding back and forth.

4 Thread the string through the hole in the cup. Tie a paper clip to the end of the string that comes out of the bottom of the cup. Use tape to attach the other end of the string to the straw.

5 Blow on the spoons to make them spin and lift the cup with the penny. Add another penny to the cup and continue to blow to lift the cup. Keep adding pennies until you can no longer make the spoons spin by blowing on them. Record the number of pennies you were able to lift.

6 Add or remove a spoon and repeat step 5.

Record

Write in your science notebook.
Use a table like this one.

Model Windmill

Number of Spoons	Number of Pennies Lifted

Explain and Conclude

1. **Compare** your data with your classmates. How did the number of spoons affect the test?

2. Describe how the energy from your blowing lifted the objects in the cup.

Think of Another Question

What else would you like to find out about wind energy? How could you find an answer to this new question?

Copenhagen, Denmark

These wind turbines convert wind energy to electrical energy.

Investigate Features on a Map

Question How can you use a **map** to identify land features?

Science Process Vocabulary

model noun

You can use a **model** to learn about things that are too big to study in real life.

compare verb

You **compare** when you tell how things are alike and different.

> I can compare different kinds of land features.

Materials

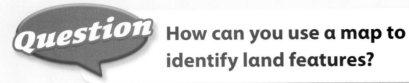

Land Features Diagram

Land Features Map

colored pencils

What to Do

1 **Observe** the land features in the photos below. Find and label each of these land features on the Land Features Diagram. Also label any rivers or lakes.

Mountain

Valley

Plain

Canyon

2 Write a description of each kind of feature in your science notebook.

What to Do, continued

3 Label the land areas on the Land Features Diagram.

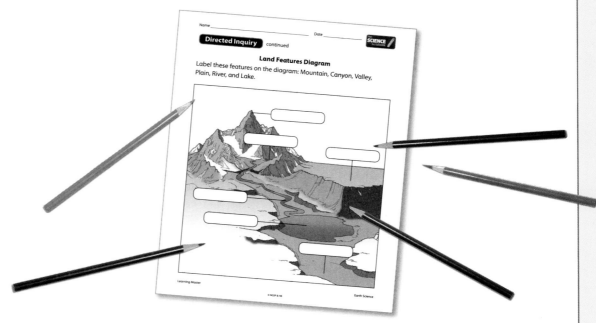

4 A map is a kind of **model** that you can use to identify land features. Observe the Land Features Map. Notice how each feature shown on the Land Features Diagram looks on the map.

5 Color all the land features on the map. Color the lakes blue and the rivers purple. Make the plains green, the mountains brown, and the canyons orange.

6 Color in the key on the map. Show the color for each kind of land feature.

Record

Write in your science notebook.
Use a table like this one.

Land Features

Land Feature	Description
Mountain	
Valley	

Explain and Conclude

1. How is the Land Features Map different from the Land Features Diagram?

2. **Compare** the different land features on the map. How does the map help you identify land features?

3. Do you think the land features formed slowly or quickly? Explain.

Think of Another Question

What else would you like to find out about maps? How could you find an answer to this new question?

Yellowstone Lake, Yellowstone National Park, Wyoming

This photo that was taken from space shows land features in Yellowstone National Park.

Investigate Glaciers

Question How can you make a model to show what happens when glaciers move?

Science Process Vocabulary

model noun

A **model** can show a process that is very slow and takes a long time.

predict verb

You **predict** when you say what you think will happen.

I predict that the model glacier will move down the hill.

Materials

safety goggles

graduated cylinder

water

cornstarch

2 plastic containers

spoon

soil

sand

rocks

book

What to Do

1 Put on your safety goggles. Use the graduated cylinder to measure 60 mL of water.

2 Put the cornstarch in one of the plastic containers. Slowly add tiny amounts of the water to the container as you stir until the mixture is as thick as toothpaste.

Don't add too much water. The mixture should not be runny.

3 The mixture is used as a **model** of a glacier. **Predict** what will happen when the glacier model flows over soil, sand, and rocks. Record your prediction.

What to Do, continued

4 With your group, decide how you will form the sand, soil, and rocks into a model of Earth's land in the other plastic container. Make your model. Then prop one end of the container on the book. Draw your model in your science notebook.

5 Slowly pour your glacier onto the highest point of your model of Earth's land. **Observe** how the model of the glacier flows and what happens to Earth's land. Record your observations.

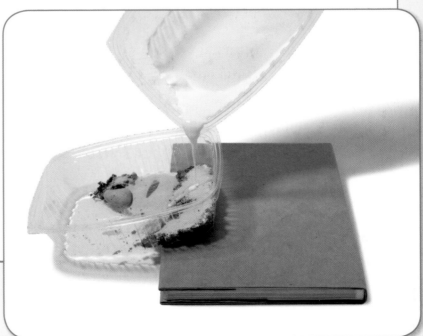

Record

Write and draw in your science notebook.
Use a table like this one.

Glacier Model Observations

What I Observed	What Happened
Model of Earth's features without glacier	
Model of Earth's features with glacier	

Explain and Conclude

1. Did your results support your **prediction?** Explain.

2. **Compare** your results with those of other groups. Did the glacier flow the same way in other **models?** What might cause differences?

3. How is your model like a real glacier? How is it different?

Think of Another Question

What else would you like to find out about glaciers? How could you find an answer to this new question?

Grey Glacier, Torres del Paine National Park, Chile

Investigate Plate Movements

Question **What are ways Earth's plates can move?**

Science Process Vocabulary

model noun

You can use a **model** to show how things work in nature.

observe verb

You **observe** when you use your senses to learn about an object or event.

I can observe the effects of plate movements.

Materials

plastic container

measuring cup

cornstarch

graduated cylinder

water

spoon

foil

tape

2 pieces of hardboard

2 pieces of foam

ruler

What to Do

1 First, you will make a **model** of the melted rock under Earth's surface. **Measure** 120 mL of cornstarch with the measuring cup and pour it into the plastic container. Slowly add water while you stir with a spoon. Add enough water to make your mixture runny but not watery.

You will use about 80 mL of water.

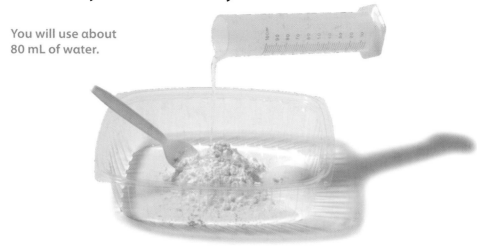

2 Tape down the foil as shown in the photo. Carefully pour the cornstarch mixture into the center of the foil.

What to Do, continued

3 The pieces of hardboard are models of large pieces of Earth's surface, or plates, found under the ocean. Put 2 pieces of hardboard on the mixture so that they touch each other. Gently push down on the pieces as you move them apart. Record your **observations** in your science notebook.

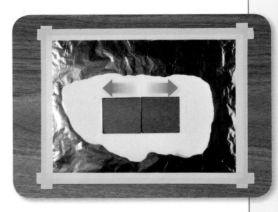

4 Put 1 hardboard piece and 1 foam piece on the mixture with 4 cm between them. The foam pieces are models of plates that are above the water. These plates make up the land. Push down on the pieces gently as you move them together. Record your observations.

5 Put 2 pieces of foam side by side on the mixture with no space between them. Push down on the pieces gently and keep them together as you move them in opposite directions. Record your observations.

Record

Write and draw in your science notebook.
Use a table like this one.

Model Plate Movements

Movement	Observations

Explain and Conclude

1. Describe three ways Earth's plates can move.

2. The melted rock under Earth's surface is very hot. What do you think would happen to real melted rock below Earth's surface if the Earth's plates moved apart as the model plates did in step 3?

3. What happened to the plates when you moved them together in step 4? How do you think Earth's surface changes when real plates move together?

Think of Another Question

What else would you like to find out about plate movements? How could you find an answer to this new question?

Great Rift Valley, Kenya, Africa

This valley formed as Earth's plates moved apart.

Investigate Landslides

Question How can you model the way earthquakes and rainfall affect landslides?

Science Process Vocabulary

measure verb

You **measure** when you find out how much or how many.

I used the graduated cylinder to measure 80 mL of water.

infer verb

You **infer** when you use what you know and what you observe to draw a conclusion.

I can infer what caused this landslide.

Materials

safety goggles

milk carton

measuring cup

sandy soil

plastic container

stopwatch

water

What to Do

1 Put on your safety goggles. You will **model** a landslide. Landslides can happen as a result of earthquakes and heavy rains.

2 Use the measuring cup to **measure** 225 mL of sandy soil. Pour the sandy soil in the closed end of the milk carton to make a hill.

3 Put the milk carton with sandy soil near the edge of a table. Have a partner hold the plastic container under the table to catch the sandy soil.

4 First, you will test how earthquakes affect landslides by modeling an earthquake. Choose whether you will shake the carton with sand gently or harder for 30 seconds. Record your choice. Then test your choice by shaking for 30 seconds. Record your **observations.**

5 Pour the sand back into the carton as you did in step 2. Now test how heavy rain affects landslides by modeling heavy rain. Put 250 mL of water in the measuring cup. Choose whether you will pour the water onto the sand slowly or quickly. Record your choice. Pour the water onto the sand. Record your observations.

6 **Compare** your results with other groups. What caused the biggest landslide?

Record

Write in your science notebook.
Use a table like this one.

	Landslides	
	Earthquakes	Heavy Rain
Choice		
Observations		

Explain and Conclude

1. How did the force of the shaking and the speed of the pouring affect the landslides?

2. Use the results of this investigation to **infer** what can cause real landslides to happen.

Think of Another Question

What else would you like to find out about landslides? How could you find an answer to this new question?

What do you think caused this landslide?

Investigate Condensation

Question How can you observe condensation and frost?

Science Process Vocabulary

measure verb

You **measure** when you find out how much or how many.

observe verb

You **observe** when you use your senses to learn about an object or event.

I observe that water condensed on this leaf.

Materials

2 index cards

2 cans

ice

salt

measuring spoon

craft stick

cool water

graduated cylinder

2 thermometers

stopwatch

What to Do

1. Write labels on 2 index cards:
Ice and Salt and **Ice and Water.**

2. Fill both cans halfway with crushed ice. Make sure that each can has about the same amount of ice.

3. Use the measuring spoon to add 2 spoonfuls of salt to 1 can. Stir the salt into the ice with the craft stick. Put the Ice and Salt index card in front of this can.

4 Do not add salt to the second can. Use the graduated cylinder to **measure** and pour 80 mL of cool water into the second can. Place the Ice and Water card in front of this can.

5 Place a thermometer in each can. Wait 2 minutes. Then **observe** the outsides of the cans and the temperatures on the thermometers. Record your observations, including the temperatures, in your science notebook.

6 Wait 3 more minutes. Then observe the outsides of the cans and the temperatures on the thermometers. Record your observations.

Record

Write in your science notebook.
Use a table like this one.

	Metal Cans			
	After 2 Min		After 3 More Min	
Can	Temperature (°C)	Observations	Temperature (°C)	Observations
Ice and salt				
Ice and water				

Explain and Conclude

1. Look for patterns in your data. What do you think caused the condensation on the cans?

2. **Compare** your observations of the 2 cans. Which can formed water on the outside faster? Which can formed frost? What do you think caused the differences?

3. How does this investigation **model** what happens in the water cycle?

Think of Another Question

What else would you like to find out about condensation? How could you find an answer to this new question?

Frost formed on this window.

Investigate Evaporation

Question How can you make water evaporate more quickly?

Science Process Vocabulary

Materials

predict verb

When you **predict,** you tell what you think will happen.

Before the investigation, I will make a prediction.

conclude verb

You **conclude** when you use data from an investigation to come up with an answer to a question.

After the investigation, I will share my conclusions with the class.

8 paper squares

cup of water

dropper

stopwatch

What to Do

1 Use the dropper to gently put 5 drops of water in the center of each paper square. Wait 30 seconds for the water to soak into the paper.

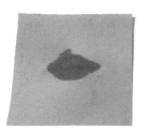

2 **Predict** which paper square will dry faster, a square you wave in the air or one that is flat on the desk. Record your predictions in your science notebook.

3 Quickly wave 1 paper square back and forth. Keep the other square flat on the desk. Use the stopwatch to record how long it takes each square to dry. Record your **data.**

What to Do, continued

4 Predict which paper square will dry faster, a crumpled one or a flat one. Record your predictions. Put 5 drops of water in the center of 2 new paper squares. Crumple 1 square into a ball. Do not crumple the other square. Record how long it takes each square to dry.

5 Predict which paper square will dry faster, a square in sunlight or a square in the shade. Put 5 drops of water on 2 more paper squares. Then place one square in sunlight and the other in the shade. Record how long it takes each square to dry.

6 With your group, think of the fastest way to make a paper square dry. Record your **plan.** Then test your idea and record your **data.**

Record

Write in your science notebook.
Use a table like this one.

Water on Paper Squares

		Prediction	Time to Dry (s)
Trial 1	Square waved in air		
	Square on desk		

Explain and Conclude

1. What can you **conclude** about what makes water evaporate more quickly?

2. **Compare** your **data** from step 6 with the data from other groups. What did you do to your paper to make it dry more quickly? What did the group with the fastest drying time do?

3. Use your observations to **infer** whether water from a lake would evaporate more quickly on a windy day or a calm day. Where would the water from the lake go?

Think of Another Question

What else would you like to find out about evaporation? How could you find an answer to this new question?

Will these items dry more quickly on a calm day or on a windy day?

Do Your Own Investigation

Choose one of these questions, or make up one of your own to do your investigation.

- How can you model the sizes of the sun, the moon, and Earth?
- Does light make some colors heat faster than others?
- How do the colors in sunlight compare to the colors in light from a flashlight?
- How does pressing down on soil affect how much water it can hold?
- How can you design a way to reuse an item instead of throwing it away?
- How do grass plants affect shoreline erosion?
- How does a building's distance from an earthquake affect how much it is damaged?
- How does the surface area of a container of water affect how quickly it evaporates?

Science Process Vocabulary

First I will take the temperature of the water. Then I will place the water in sunlight.

plan noun

When you make a **plan** to answer a question, you list the materials and steps you need to take.

Open Inquiry Checklist

Here is a checklist you can use when you investigate.

- ☐ Choose a **question** or make up one of your own.

- ☐ Gather the materials you will use.

- ☐ If needed, make a **hypothesis** or a **prediction.**

- ☐ If needed, identify, manipulate, and control **variables.**

- ☐ Make a **plan** for your **investigation.**

- ☐ Carry out your plan.

- ☐ Collect and record **data. Analyze** your data.

- ☐ Explain and **share** your results.

- ☐ Tell what you **conclude.**

- ☐ Think of another question.

Write Like a Scientist

Write About an Investigation

Colors and Sunlight

After Sharee stood outside on a cold day, she began to wonder about colors and sunlight. She asked herself, "Will I feel warm faster if I wear certain colors?" She decided to do an investigation. Here is what she thought about:

- Sharee decided that using clothing to find the answer would be too difficult.
- She thought she could measure how fast materials heat by making cones out of different colors of paper. She would place the cones over ice cubes to see which ice cube melted fastest.

- She didn't know if the sun would be shining when she did her investigation so she decided to use a lamp to represent the sun.

Model

Question
Does light make some colors heat faster than others?

Choose a question that can be answered using materials that are safe and easy for you to get.

Materials

newspaper

drawing compass

4 sheets of construction paper: white,
    black, yellow, and blue

scissors

tape

4 ice cubes

desk lamp

stopwatch

List all the materials you will need.

Your Investigation

Now it's your turn to do your investigation and write about it. Write about the following checklist items in your science notebook.

☐ Choose a question or make up one of your own.

☐ Gather the materials you will use.

Model

My Hypothesis

If I cover 4 ice cubes with paper cones of different colors, then some ice cubes will melt faster than others. The ice cubes that melt faster will have heated faster.

You can use "If…, then…." statements to make your hypothesis clear.

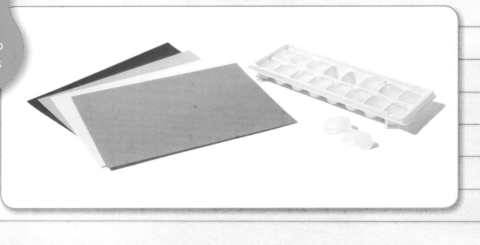

 **Your Investigation**

☐ **If needed, make a hypothesis or prediction.**

Write your hypothesis or prediction in your science notebook.

Model

Variable I Will Change

I will change the color of the paper cone over the ice cube. One ice cube will be covered by white paper, one by black paper, one by yellow paper, and one by blue paper.

Variable I Will Observe or Measure

I will measure how much ice is left for each cube after 20 minutes.

Variables I Will Keep the Same

Everything else will be the same. All the ice cubes and the paper cones will be the same size. All the ice cubes will be the same distance from the light.

Answer these three questions:
1. What one thing will I change?
2. What will I observe or measure?
3. What things will I keep the same?

Your Investigation

☐ If needed, identify, manipulate, and control variables.

Write about the variables for your investigation.

Model

My Plan

1. Spread newspaper on a table to keep the area dry.
2. Use the drawing compass to draw a 15 cm circle on each sheet of construction paper. Cut out the circles.
3. Make a cut from the edge of each circle to the center of the circle. Shape each circle into a cone. Tape the edges.
4. Place 4 ice cubes on the newspaper under a lamp.
5. Cover each cube with a different color cone.
6. Turn on the lamp. Observe the ice cubes every 5 minutes for 20 minutes to see how much they have melted.

Write detailed plans. Another student should be able to do your investigation without asking any questions.

 **Your Investigation**

☐ **Make a plan for your investigation.**

Write the steps for your plan.

Model

I carried out all six steps of my plan.

Your Investigation

☐ **Carry out your plan.**

Be sure to follow your plan carefully.

Make a note if you need to change your plan in any way. Sharee did not have to make any changes.

Model

Data (My Observations)

Observations of Ice Cubes

Paper Color	Amount of Ice Left			
	After 5 Minutes	After 10 Minutes	After 15 Minutes	After 20 Minutes
White				
Black				
Yellow				
Blue				

> Use a table to organize your data.

My Analysis

At the start, all the ice cubes were the same size.

After 20 minutes, the ice cube under the black cone had melted more than the other ice cubes.

> Explain what happened, based on the data you collected.

 **Your Investigation**

☐ **Collect and record data. Analyze your data.**

Collect and record your data, and then write your analysis.

How I Shared My Results

First, I explained that I measured how fast ice melts as a way to find out how fast different colors heat up. Next, I drew pictures to show how much ice was left under each color cone. Then, I shared my data.

My Conclusion

The ice cube under the black cone melted faster than the ice cube under the other cones. That meant that the black construction paper heated up faster than the other construction paper. The results of the investigation support my hypothesis.

Tell what you conclude and what evidence you have for your conclusion.

Another Question

I wonder if some surfaces heat faster in sunlight. Do dull surfaces heat faster than shiny surfaces?

my **SCIENCE** notebook / **Your Investigation**

Investigations often lead to new questions for inquiry.

- ☐ **Explain and share your results.**
- ☐ **Tell what you conclude.**
- ☐ **Think of another question.**

Science and Technology

Using Solar Energy

Solar energy is energy from the sun. It is Earth's most plentiful energy source. Solar energy is a renewable energy source. A renewable energy source is a resource that will not run out.

Energy from the sun can be changed into other forms of energy, such as heat and electricity.

Solar energy can be changed into electricity using a solar cell. Most calculators contain a small solar cell.

Solar cell

The small, flat solar cell can change solar energy into electricity to run a calculator.

Solar Power Plants Some sunny, warm states are developing solar energy power plants. These power plants use solar energy to make electricity.

This solar farm in Arcadia, Florida, can make electric power for many homes.

Solar cells

Solar Water Heating Solar energy is also used to heat water in people's homes. The simplest solar water heating systems are called passive solar water heaters. This type of solar heater is most common in places without long periods of freezing temperatures, such as in much of the southern United States.

Solar energy collector
(water is heated)

Cold water in

Hot water out

Passive solar water heaters are made of two main parts: a solar collector and a tank to store the heated water.

Tank to store hot water

People use a solar cooker to cook rice in a village in Zambia, Africa.

Solar Ovens You may know that sunlight can make objects warm, or even hot. Hundreds of years ago, people used sunlight to start fires. Then about 200 years ago, a scientist built a solar oven.

Today different kinds of solar ovens and solar cookers are used all over the world. Solar cookers are cheap to make and easy to use. People can easily carry them from one place to another. Most importantly, solar cookers can heat food and water hot enough to make them safe to eat and drink.

SUMMARIZE

What Did You Find Out?

1. What is solar energy?

2. Why is solar energy a good source of energy?

3. How is solar energy used?

Observe Solar Energy

You can observe how the sun's energy heats objects. Follow the steps below using 3 smooth black rocks.

1. Build a "house" for 1 of the rocks. Place a small box inside a larger box. Use paper towels to fill the spaces between the boxes.

2. Put the rock inside the house and cover the opening with plastic wrap and a rubber band. Place the house in a sunny spot so that the sunlight shines directly on the rock.

3. Place the other 2 rocks in the sunlight next to the house. Cover 1 rock with a clear plastic cup.

4. After 30 minutes, feel each rock. Which rock feels the warmest? Explain why you think this is so.

Science in a Snap!

CHAPTER 1

Science in a Snap! Water Predictions

Observe the 2 containers. **Predict** which container has the most water. Then pour the water from 1 container into a measuring cup. **Measure** and record the amount of water in the cup. Pour the water back into its container. Repeat for the other container. Was your prediction supported? Explain.

CHAPTER
2

Science in a Snap! Classify Solids, Liquids, and Gases

Write **Solid** on a sticky note. Look around your classroom and find an object that is a solid. Put the sticky note on the object. Then write **Liquid** and **Gas** on 2 more sticky notes. Find and label a liquid and a gas in your classroom. How did you know to **classify** matter as a solid, a liquid, or a gas?

Solid

Liquid

Gas

CHAPTER
3

Science in a Snap! Compare Motion

Place a book on a desk. Push the book and **observe** its motion. Did the book move toward you or away from you? How else could you describe the book's motion? Next, roll a ball across the desk to a partner. Describe the motion of the ball. Have your partner roll the ball back. **Compare** the direction of the ball's motion when you rolled it and when your partner rolled it. How was the motion of the ball different from the motion of the book?

Science in a Snap! Make a Sound

Put your fingers on your throat. Make a low humming sound and **observe** what you feel under your fingertips. Now make a high humming sound and observe what you feel. Stop humming and observe what happens. Now start humming again, this time more loudly. What do you feel? **Infer** what causes the humming sound you hear.

CHAPTER 5

Science in a **Snap!** Observe Heat

Cut out squares of white construction paper, black construction paper, foil, tissue, and plastic wrap. Place the materials in front of a lamp. Turn on the lamp. Wait 10 minutes, and then turn off the lamp. **Compare** how warm each material feels. Why might some materials feel warmer than others?

Investigate Physical Properties

Question How can you use tools to investigate physical properties?

Science Process Vocabulary

observe verb

You **observe** when you use your senses to learn about objects or events.

predict verb

You **predict** when you say what you think will happen.

> I predict that the shape of the cloth will change when I fold it.

Materials

burlap

hand lens

microscope

scissors

wool

sandpaper

paper towel

silk

What to Do

1 **Observe** the physical properties of the burlap with just your eyes. Look at properties such as its color and whether it looks rough or smooth. Record your observations in your science notebook.

2 Observe the properties of the burlap with the hand lens. Record your observations.

3 **Predict** how your observations will change if you observe the burlap with a microscope. Record your prediction. Then observe the burlap with the microscope. Record your observations.

Turn this knob to focus.

4 Predict how the burlap's physical properties will change if you cut it into very small pieces. Record your prediction.

5 Cut 4 very small pieces from the burlap. Observe the pieces with just your eyes. Then use the hand lens and the microscope. Record your observations.

6 Repeat steps 1–5 with the other materials.

Record

Write in your science notebook.
Use tables like these.

Predictions

Material	How will my observations change when I use the microscope?	How will the material's physical properties change when I cut it?
Burlap		

Observations of Physical Properties

Material	Before Cutting			After Cutting		
	Just Eyes	Hand Lens	Microscope	Just Eyes	Hand Lens	Microscope
Burlap						

Explain and Conclude

1. Did your results support your **predictions?** Explain.

2. How did tools help you to better **observe** the objects?

3. **Compare** the physical properties of the large pieces of the materials with the physical properties of the very small pieces.

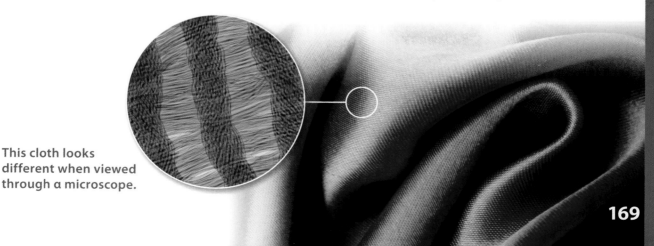

This cloth looks different when viewed through a microscope.

Investigate Volume and Mass

Question How can you compare the volume and mass of solid and liquid objects?

Science Process Vocabulary

measure verb

You can use tools to **measure** mass and volume.

predict verb

You **predict** when you say what you think your results will be in an investigation.

I predict that the marble has more mass than the rock.

Materials

graduated cylinder

water

marble

rock

cup

balance

gram masses

What to Do

1 Put 20 mL of water into the graduated cylinder. Add the marble. **Measure** the volume of the water and the marble. Record your **data** in your science notebook.

2 To find the volume of the marble, subtract the volume of the water from the volume of the water and the marble. Record your data.

3 Repeat steps 1–2 with the rock.

4 Now you will measure mass. Which do you think has the most mass—the marble, the rock, or 20 mL of water? Write your **prediction.**

5. Use the balance to find the mass of the cup. Record your data.

6. Place the marble in the cup and find the mass of the cup and marble. Record your data. To find the mass of the marble, subtract the mass of the cup from the mass of the cup and marble. Record your data.

7. Repeat step 6 with the rock and with 20 mL of water.

Record

Write in your science notebook.
Use tables like these.

Volume

Object	Volume of Water (mL)	Volume of Water and Object (mL)	Volume of Object (mL)
Marble	20		
Rock	20		

Mass

Object	Mass of Cup (g)	Mass of Cup and Object (g)	Mass of Object (g)
Marble			
Rock			
Water			

Explain and Conclude

1. **Compare** the volume of the water, marble, and rock. Which had the most volume?

2. Which object had the most mass? Is that what you **predicted?** Explain.

3. **Share** your results with others. Explain any differences.

Think of Another Question

What else would you like to find out about comparing the volume and mass of solids and liquids? How could you find an answer to this new question?

Early balance

173

Investigate Water and Temperature

Question What happens to water as the temperature changes?

Science Process Vocabulary

Materials

observe verb

When you **observe,** you use your senses to learn about objects or events.

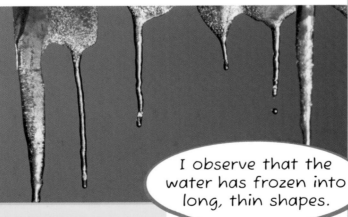

I observe that the water has frozen into long, thin shapes.

compare verb

When you **compare,** you tell how objects or events are alike and different.

The icicles are all clear and cold, but they have different shapes.

2 resealable bags

tape

graduated cylinder

water

What to Do

1 Label the plastic bags **Bag 1** and **Bag 2.** Use the graduated cylinder to **measure** 100 mL of water. Pour the water into one bag. Seal the bag. Repeat with the other bag.

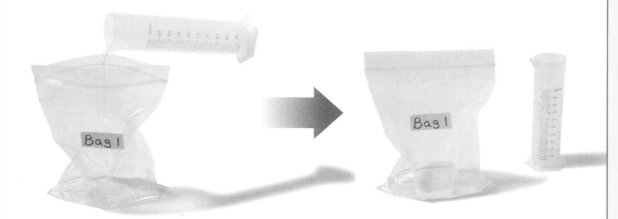

2 Carefully place both bags in the freezer. **Predict** what will happen to the water in the bags. Record your prediction in your science notebook.

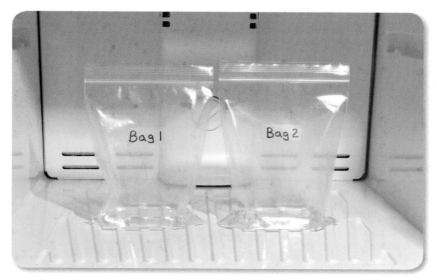

What to Do, continued

3 The next day, take the bags out of the freezer. Turn the bags in different directions. **Observe** the shape of the water. Record your observations.

4 Put the bags on your desk. Wait 30 minutes. Turn the bags in different directions. Record your observations.

5 Put the bags in sunlight. Open Bag 1, being careful not to spill the water. Keep Bag 2 sealed. Predict what will happen to the water in both bags after 3 days.

6 Observe the bags every day for the next 3 days. Record your observations.

Record

Write or draw in your science notebook.
Use a table like this one.

Observations of Bags with Water		
	Bag 1	Bag 2
After 1 day in the freezer		
After 30 minutes on the desk		

Explain and Conclude

1. Did your results support your **predictions?** Explain.

2. **Compare** your results with the results of other groups. What patterns do you see?

3. Use your results from steps 2–6 to **conclude** what happens to water when the temperature goes down. What happens to frozen water when the temperature goes up?

Think of Another Question

What else would you like to find out about what happens to water as the temperature changes? How could you find an answer to this new question?

What will happen to the water on this tomato as time passes?

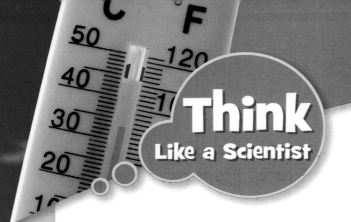

Math in Science

Measuring Temperature

Scientists use thermometers to measure temperature. Thermometers help scientists make better observations. Scientists can get a more exact temperature with a thermometer than by just feeling how hot or cold something is. Also, scientists can use thermometers to measure the temperatures of objects that are too hot or too cold to touch.

Digital thermometer

Liquid crystal thermometer

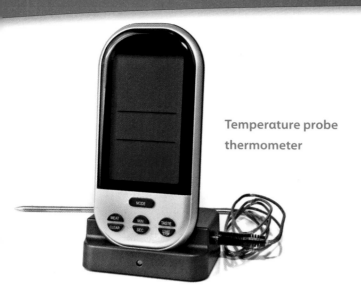

Temperature probe thermometer

The simplest thermometer is made up of a clear tube that contains a liquid. As the temperature gets warmer, the liquid rises in the tube. When the temperature gets cooler, the liquid moves lower in the tube.

Temperature Scales Look at the scale on both sides of the thermometer. One scale shows the temperature in degrees Fahrenheit (°F). You may have heard the temperature given in degrees Fahrenheit during a television weather report.

The other scale shows the temperature in degrees Celsius (°C). Scientists usually measure temperature in degrees Celsius.

Using a Thermometer

When reading a thermometer, follow the steps below. For steps 2 and 3, use the thermometer in the photograph to practice.

1 Place the thermometer on or in the material for which you want to find the temperature. Wait about 1 minute for the thermometer to complete its reading.

2 Put the top of your finger at the top of the red liquid.

3 Slide your finger to the right to read the Celsius temperature. Slide your finger to the left to read the Fahrenheit temperature.

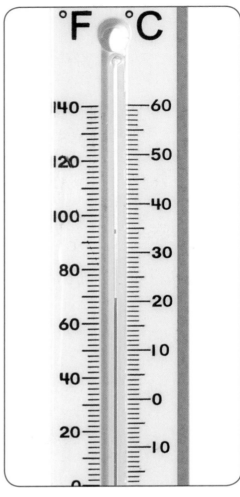

When you measure temperature, make sure you understand the number of degrees each mark shows.

SUMMARIZE
What Did You Find Out?

1 What are two reasons why scientists use thermometers to measure temperature?

2 What two scales are used to measure temperature? Which do scientists usually use?

Measure the Temperature of Water

Practice using a thermometer. Follow these steps:

1 Fill a plastic cup half full of water.

2 Place a thermometer in the water. Wait 1 minute.

3 Find the temperature of the water on both the Fahrenheit and Celsius scales. Record your measurements.

4 Switch cups with a partner. Measure the temperature in your partner's cup of water.

5 Compare your temperature readings with your partner's readings. Explain why there might be differences between your measurements and your partner's measurements.

Investigate Melting

Question How does heating and cooling affect the properties of different materials?

Science Process Vocabulary

experiment noun

In an **experiment,** you change only one variable. You measure or observe another variable. You control other variables so they stay the same.

Some variables in this experiment are the amount of liquid, kind of liquid, and kind of container.

variable noun

A **variable** is something that can change in an experiment.

Materials

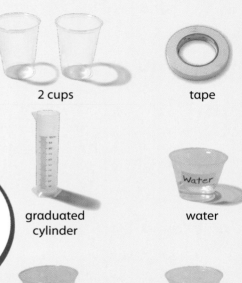

2 cups

tape

graduated cylinder

water

salt water

vinegar

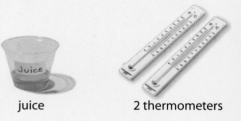

juice

2 thermometers

stopwatch

Do an Experiment

Write your plan in your science notebook.

Make a Hypothesis

In this experiment, you will investigate how heat energy from sunlight can affect the properties of two frozen materials. You will freeze two liquids. One liquid will be water. You will choose the second liquid. Then you will place the frozen materials in sunlight. What will happen to the temperature and properties of the materials as they sit in sunlight? Write your **hypothesis.**

Identify, Manipulate, and Control Variables

Which variable will you change?
Which variable will you observe or measure?
Which variables will you keep the same?

What to Do

1 Label 1 cup **Water.** Then use the graduated cylinder to **measure** 50 mL of water. Pour the water into the cup.

What to Do, continued

2 Choose a different liquid. Label a cup with the name of the liquid. Put 50 mL of the liquid into the cup.

3 Put a thermometer in each cup. **Measure** the temperature of each liquid and **observe** their properties. Record your **data** in your science notebook.

4 Put the cups with liquid in a freezer. The next day, take the cups out of the freezer. Observe the properties and record the temperature of each material.

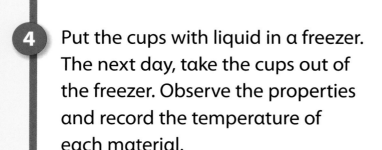

5 Place the cups in sunlight. Every 10 minutes, observe each cup and measure the temperature of each liquid. Record your observations and the temperatures. Observe the materials every 10 minutes until all the materials have melted.

Record

Write in your science notebook.
Use a table like this one.

Liquids in Cups

	Water		My Choice: _____	
	Temperature	Observations	Temperature	Observations
Before freezing				
After freezing				
After 10 min in sunlight				
After 20 min in sunlight				

Explain and Conclude

1. What happened to the temperature of the liquids in the freezer? What happened to the temperature of the materials in sunlight?

2. How did the properties of the materials change as their temperatures changed? Do these results support your **hypothesis?** Explain.

3. **Compare** the results of all groups. What patterns do you **observe?**

Think of Another Question

What else would you like to find out about changes in temperature and properties of materials? How could you find an answer to this new question?

The heat from sunlight melts the ice.

Investigate Forces and Motion

Question How do forces affect motion?

Science Process Vocabulary

measure verb

When you **measure,** you find out how much or how many.

compare verb

You **compare** when you tell how things are alike and different.

Materials

3 books

table tennis ball

meterstick

tape

fabric

I can compare how far the ball moves with a strong push and a soft push.

186

What to Do

1 Make a low ramp using 1 thin book and 1 thick book. You will roll the table tennis ball down the ramp.

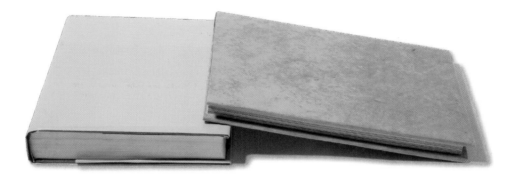

2 Put the ball on the top of the ramp. When a ball rolls down a low ramp, it moves with little force. **Predict** how far the ball will move when you let it go down the ramp. Test your prediction. Release the ball and use the meterstick to **measure** how far it moves after it leaves the ramp. Record your **data** in your science notebook.

What to Do, continued

3 Add another thick book to make a high ramp. When a ball rolls down a high ramp, it rolls with greater force. Predict how far the ball will move when you let it go down the 2-book ramp. Test your prediction. Release the ball and use the meterstick to measure how far it moves. Record your data.

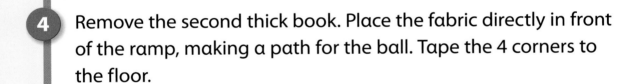

4 Remove the second thick book. Place the fabric directly in front of the ramp, making a path for the ball. Tape the 4 corners to the floor.

5 Repeat steps 2 and 3 with the fabric. Record your predictions and measurements.

Record

Write in your science notebook.
Use a table like this one.

Motion of Ball

Trial	Prediction (cm)	How Far Ball Moves (cm)
Low ramp on floor		
High ramp on floor		
Low ramp on fabric		
High ramp on fabric		

Explain and Conclude

1. Did your results support your **predictions?** Explain.

2. **Compare** how the ball moved when you rolled it down the low and the high ramps. What do you think caused the difference?

3. Compare how the ball moved on the floor and on the fabric. What do you think caused the difference?

Think of Another Question

What else would you like to find out about forces and motion? How could you find an answer to this new question?

The roughness of the sand slows down the ball.

Investigate Motion and Position

 **Question** How can you describe the motion of an object by observing its position?

Science Process Vocabulary

plan noun

You make a **plan** to answer a question. You list the materials and steps you will take.

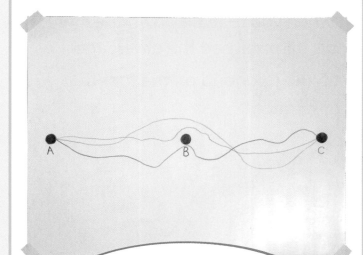

I made a plan to do three trials. I will observe and draw the motion of the ball during each trial.

Materials

poster board

tape

3 markers

ball

2 straws

spoon

string

ruler

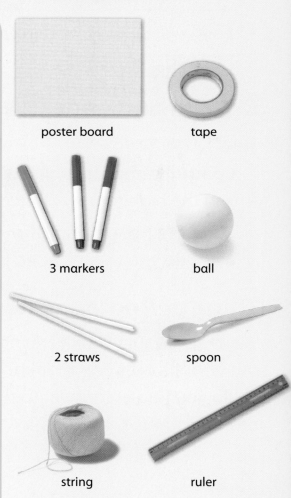

What to Do

1 Tape the poster board to a flat surface. Make 3 circles on the poster board as shown in the photo. Label the circles **A, B,** and **C.**

2 Make a **plan** to move the ball from circle A to circle C. You must move the ball as close to circle B as possible, but the ball may not roll over it. You can choose 2 straws, a spoon, or string to move the ball. Record your plan in your science notebook.

3 Place the ball at circle A. Test your plan. As you move the ball, have a partner use a marker to trace the ball's path on the poster board.

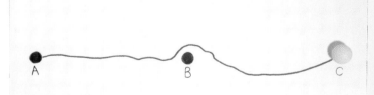

4 Repeat step 3 two more times. Use a different color marker each time to trace the ball's path.

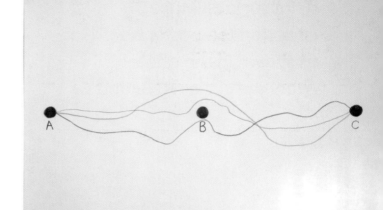

5 Use a ruler to **measure** the distance from each line to circle B. Record your **data.**

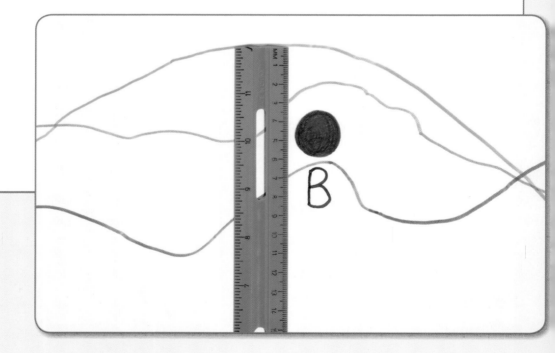

Record

Write in your science notebook.
Use a table like this one.

Path of Ball

Trial	Distance from Circle B (cm)
1	
2	
3	

Explain and Conclude

1. How did the ball's position change in relation to circle A? How did it change in relation to circle C?

2. **Compare** your **data.** In which trial did the ball come closest to circle B? Describe the path of the ball in that trial. Describe the ball's position in relation to circle B.

3. **Conclude** how you can use position to describe motion.

Think of Another Question

What else would you like to find out about motion and position? How could you find an answer to this new question?

The car is moving toward the tree.

Investigate Energy of Motion

Question How does adding more washers affect the motion of a pendulum?

Science Process Vocabulary

count verb

I counted 6 pendulums.

When you **count,** you tell the number of something.

compare verb

When you **compare,** you tell how objects or events are alike and different.

I can compare the number of swings each pendulum makes in 1 minute.

Materials

safety goggles

string with paper clip hook

tape

metal washers

ruler

stopwatch

What to Do

1 Put on your safety goggles. Tape the free end of the string to the edge of a table. Put a metal washer on the paper clip hook. The metal washer should be close to the floor but not touching it. You have made a pendulum.

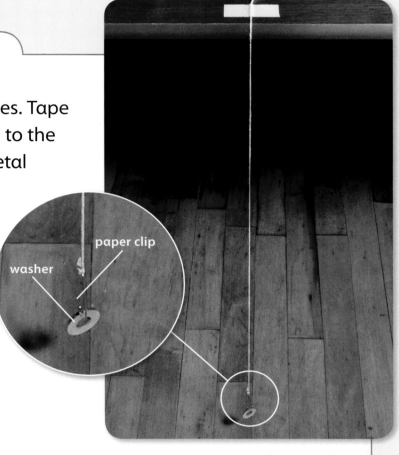

washer

paper clip

2 Pull the pendulum back 10 cm, and then let it go. Use the stopwatch to **count** the number of swings the pendulum makes in 1 minute. One swing is when the pendulum moves back and forth one time. Record your **data** in your science notebook.

3 Add 2 more washers to the paper clip hook. How do you think adding more washers will affect the number of pendulum swings in 1 minute? Record your **prediction.** Repeat step 2. Be sure to pull the pendulum back 10 cm before letting it go.

4 Add 2 more washers to the paper clip hook. How do you think adding even more washers will affect the number of pendulum swings in 1 minute? Record your prediction. Repeat step 2.

5 Use your data to make a bar graph to **compare** the pendulum swings. Look for a **pattern** in the graph.

Record

Write in your science notebook. Use your data to make a bar graph like the one shown here.

Pendulum Swings

Number of Washers	Prediction	Number of Swings
1		
3		
5		

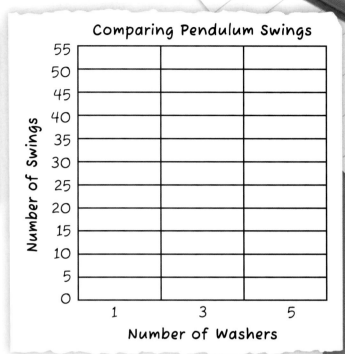

Explain and Conclude

1. Did your results support your **predictions?** Explain.

2. **Compare** the number of swings the pendulum made with 1, 3, and 5 washers. How did increasing the number of washers affect the motion of the pendulum?

3. Describe the motion of the pendulum. Where in its swing does the pendulum move fastest? Where does it move slowest?

Think of Another Question

What else would you like to find out about how adding more washers can affect the swing of a pendulum? How could you find an answer to this new question?

Investigate Vibrations and Sound

Question **How does the length of a tuning fork affect the pitch of the sound it makes?**

Science Process Vocabulary

observe verb

When you **observe,** you use one or more of your senses to learn about an object or event.

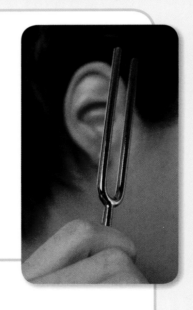

infer verb

You **infer** when you use what you know and what you observe to draw a conclusion.

I can infer how the length of the tuning fork affects the sound it makes.

Materials

tuning forks

ruler

plastic cup

plastic wrap

rubber band

salt

eraser

What to Do

1 Choose a set of tuning forks to test. **Measure** the length of the tines on each tuning fork. Record your measurements in your science notebook.

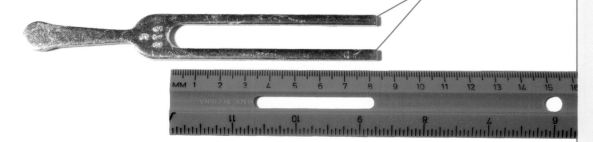

2 Stretch a piece of plastic wrap tightly across the top of the plastic cup. Use the rubber band to hold the plastic wrap in place. Sprinkle a small amount of salt on the plastic wrap.

What to Do, continued

3 Place the eraser on the desk. Hit one of the tines of the shorter tuning fork against the eraser. Touch one of the tines to the plastic wrap. **Observe** what happens to the salt. Repeat with the longer tuning fork. Record your observations in your science notebook.

4 Hit one of the tines of the shorter tuning fork against the eraser. Hold the tines near, but not touching, your ear. Listen to the sound. Repeat with the longer tuning fork. Which tuning fork has a high pitch? Which tuning fork has a low pitch? Record your observations.

5 **Compare** your observations with the observations of other groups. What patterns do you notice about the length of the tuning forks and the sounds they make?

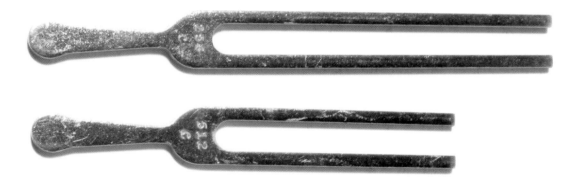

Record

Write in your science notebook.
Use a table like this one.

Tuning Forks

Tuning fork Length (cm)	What Happened to the Salt When You Touched the Tine to the Plastic Wrap?	Did It Make a Sound With a High or Low Pitch?

Explain and Conclude

1. What happened to the salt when you touched the tuning forks to the plastic wrap? Explain why you think this happened.

2. **Compare** the sounds made by the tuning forks. How did the length of the tuning fork affect the pitch of the sound it made?

3. When an object vibrates faster, it makes a sound with a higher pitch than an object that is vibrating more slowly. Use your **observations** to **infer** which of your tuning forks vibrated faster.

Think of Another Question

What else would you like to find out about vibrations and sound? How could you find an answer to this new question?

This musician is using a tuning fork to tune the piano.

201

Directed Inquiry

Investigate Light and Heat

Question What happens to an object's temperature when light shines on it?

Science Process Vocabulary

measure verb

When you **measure,** you find out how much or how many.

conclude verb

When you **conclude,** you use information, or data, from an investigation to come up with a decision or answer.

I conclude that the temperature of the sand is higher than the temperature of the water.

Materials

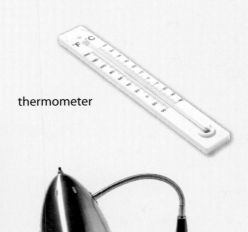

thermometer

lamp

stopwatch

What to Do

1 Put the thermometer on your desk. Place the lamp over the thermometer. Do not turn on the lamp. Wait 5 minutes. Then record the temperature on the thermometer in your science notebook as the Start temperature.

2 **Predict** what will happen to the temperature if you turn on the lamp. Record your prediction.

3 Turn on the lamp. Wait 5 minutes. **Measure** and record the temperature on the thermometer.

4 Predict what will happen to the temperature if you leave the lamp on for 5 more minutes. Record your prediction. Then wait 5 minutes. Record the temperature on the thermometer.

5 Predict what will happen to the temperature if you turn the lamp off. Record your prediction. Turn off the lamp. Move the thermometer to a place on your desk that has not been heated by the lamp. Wait 5 minutes. Record the temperature on the thermometer.

6 Predict what will happen to the temperature if you leave the lamp off for 5 more minutes. Record your prediction. Wait 5 more minutes. Then, record the temperature on the thermometer again.

Record

Write in your science notebook.
Use a table like this one.

Light and Temperature

	Temperature (°C)	Predictions
Start		What will happen to the temperature if you turn on the lamp?

Explain and Conclude

1. Did your **observations** support your **predictions?** Explain.

2. What happened to the temperature as the lamp shined on the thermometer longer? What happened after you turned off the lamp?

3. What can you **conclude** about what can happen to the temperature of an object when light shines on it?

Think of Another Question

What else would you like to find out about what happens to an object's temperature when light shines on it? How could you find an answer to this new question?

In some restaurants, lights are used to keep food warm.

Investigate Light and Objects

Question **What happens to light when it shines on different objects?**

Science Process Vocabulary

Materials

compare verb

You **compare** when you say how things are alike and how they are different.

You can compare the results of your investigation with another group's results.

When I compare the information, I see that both groups had the same results.

white paper

tape

meterstick

flashlight

mirror

foil

black paper

cloth

paper towel

What to Do

1 Tape a piece of white paper to the wall at eye level. **Measure** a distance of 50 cm from the paper. Put a piece of tape at that distance.

2 Stand at the piece of tape. Turn on the flashlight. Shine the flashlight on the white paper. **Observe** the light on the paper. Record your observations in your science notebook. Turn off the flashlight.

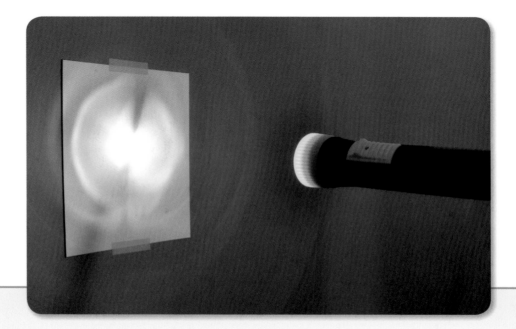

What to Do, continued

3 Select 3 objects. **Predict** what will happen to the light on the white paper when you put each object between the flashlight and the paper. Record your predictions.

4 Have a partner hold 1 object in front of you. Hold the flashlight behind the object, pointing it toward the white paper.

5 Turn on the flashlight. Observe the amount of light on the white paper and the object. Record your observations. Repeat with the other 2 objects.

Record

Write in your science notebook.
Use a table like this one.

Light on Paper

Material Between Flashlight and Paper	Predictions	Observations	
	On the white paper	On the white paper	On the object
None			

Explain and Conclude

1. **Compare** how bright the light on the white paper was when you shined the flashlight at the 3 different objects.

2. Were you able to see the same amount of light on all 3 objects in step 5? Why do you think that is so?

3. Based on the results of your **investigation,** what can you **infer** about what happens to light when it shines on different objects?

Think of Another Question

What else would you like to find out about what happens to light when it shines on different objects? How could you find an answer to this new question?

Do Your Own Investigation

Question **Choose one of these questions, or make up one of your own to do your investigation.**

- How do the masses of different pennies compare?
- How can you make water evaporate more quickly?
- How does the length of a ramp affect the distance traveled by a toy car?
- How does changing the tightness of a piece of fishing line affect the pitch of the sound it makes?
- How can you line up mirrors to make light reflect on a certain spot?

Science Process Vocabulary

question noun

You ask a **question** to find out about something.

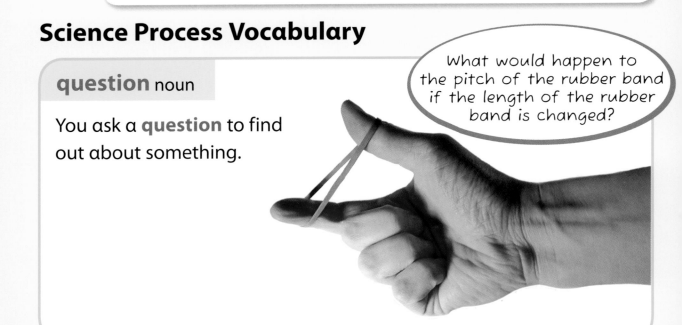

What would happen to the pitch of the rubber band if the length of the rubber band is changed?

Open Inquiry Checklist

Here is a checklist you can use when you investigate.

- ☐ Choose a **question** or make up one of your own.

- ☐ Gather the materials you will use.

- ☐ If needed, make a **hypothesis** or a **prediction.**

- ☐ If needed, identify, manipulate, and control **variables.**

- ☐ Make a **plan** for your **investigation.**

- ☐ Carry out your plan.

- ☐ Collect and record **data. Analyze** your data.

- ☐ Explain and **share** your results.

- ☐ Tell what you **conclude.**

- ☐ Think of another question.

A siren has a high pitch.

Write Like a Scientist

Write About an Investigation

Vibrations and Pitch

The following pages show how one student, Neal, wrote about an investigation. As Neal watched a musician tune his guitar, he wondered how changing the tightness of the guitar strings affected the sound the guitar made. He decided to investigate. Here is what he thought about to get started:

- Neal thought that changing the tightness of the guitar string would change the pitch of the sound the string made when plucked. He decided to do an investigation that showed how the tightness of a string affects pitch.
- He wanted to use simple materials. He would base his question and steps on the materials he would use.
- He decided to make a model guitar with a pegboard and fishing line.
- He would attach the piece of fishing line to the pegboard with bolts and washers. He would change the tightness of the fishing line.

Model

Question

How does changing the tightness of a piece of fishing line affect the pitch of the sound it makes?

Choose a question that can be answered using materials that are safe and easy to obtain.

Materials

pegboard

fishing line

2 bolts

2 nuts

2 washers

List exact amounts of materials when needed.

 Your Investigation

Now it's your turn to do your investigation and write about it. Write about the following checklist items in your science notebook.

☐ Choose a question or make up one of your own.

☐ Gather the materials you will use.

Model

My Hypothesis

If I tighten the fishing line, then the sound it makes will have a higher pitch.

> You can use "If…, then…." statements to make your prediction clear.

 **Your Investigation**

☐ **If needed, make a hypothesis or prediction.**

Write your hypothesis or prediction in your science notebook.

Model

Variable I Will Change

I will change how tight the fishing line is stretched over the pegboard.

Variable I Will Observe

I will observe the pitch of the sound the fishing line makes when it is plucked.

Variables I Will Keep the Same

Everything else will be the same. I will pluck the fishing line with the same amount of force each time. The length of the fishing line will be the same each time.

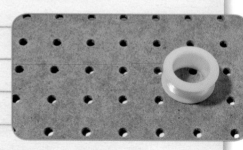

 **Your Investigation**

☐ If needed, identify, manipulate, and control variables.

Answer these three questions:
1. What one thing will I change?
2. What will I observe or measure?
3. What things will I keep the same?

Write about the variables for your investigation.

Model

My Plan

1. Place the washer on the bolt. Place the washer and bolt through a hole at one end of the pegboard. The top of the bolt should be on top of the pegboard.

2. Tighten the nut on the bolt.

3. Tie a piece of fishing line around the bolt. Pull the fishing line as tight as possible.

4. Tie the other end of the fishing line to another bolt, washer, and nut on the other end of the pegboard.

5. Turn the bolt to tighten the fishing line. Pluck the fishing line and listen to its pitch.

6. Tighten the fishing line again. Pluck it to see how the pitch changes.

7. Repeat step 6.

Write detailed plans. Another student should be able to repeat your investigation without asking any questions.

Your Investigation

☐ **Make a plan for your investigation.**

Write the steps for your plan.

Model

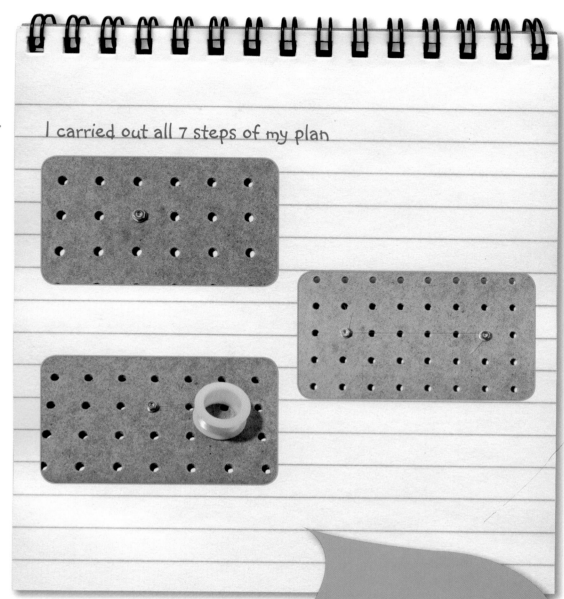

I carried out all 7 steps of my plan

You might make a note if you needed to adjust your plan in any way. Neal decided that he would add 2 more pieces of fishing line to his model the next time he did the experiment.

 **Your Investigation**

☐ **Carry out your plan.**

Be sure to follow your plan carefully.

Model

Data (My Observations)

Pitch of String

	Tightness of Fishing Line	Pitch of Sound
Trial 1	loose	low
Trial 2	tighter	higher
Trial 3	tightest	highest

My Analysis

The fishing line made a sound with a higher pitch when it was stretched tighter.

> Explain what happened based on the data you collected.

Your Investigation

☐ **Collect and record data. Analyze your data.**

Collect and record your data, and then write your analysis.

Model

Scientists often share results by demonstrating something.

How I Shared My Results

I played my model guitar for the class. I demonstrated what happened to the pitch of the fishing line when I changed its tightness. Then I shared my conclusions.

My Conclusion

Tell what you conclude and what evidence you have for your conclusion.

As the fishing line is tightened, the pitch becomes higher.

Another Question

I wonder what would happen if I used strings of different thicknesses. Would the pitch change the same way?

Investigations often lead to new questions for Inquiry.

Your Investigation

- ☐ Explain and share your results.
- ☐ Tell what you conclude.
- ☐ Think of another question.

How
Scientists Work

Using Observations to Evaluate Explanations

Scientists test their ideas by doing investigations. They make observations and gather information about the natural world. Then they analyze and interpret the information to explain their ideas.

Making Observations and Collecting Data When scientists investigate, they gather information, called data. Data can be gathered by seeing, hearing, smelling, or feeling something.

Recording and Organizing Data Scientists might record data as written descriptions of what they observe. For example, a scientist might describe the color of an object or its motion. He or she might describe how an animal acts.

Other times data might be recorded as drawings or photographs. If scientists use measuring tools to make observations, their data might be numbers.

This data was recorded by the famous scientist Albert Einstein.

Analyzing Data Then scientists analyze their data. They look for patterns. They decide whether the data support their ideas and explanations.

Sometimes the results of an investigation may not support a scientist's ideas. Then the scientist might ask questions such as:

• Did I make mistakes?

• Do I need to collect more data?

• What else do I need to know?

• How can I change my ideas so that they are supported by the new data?

SUMMARIZE

What Did You Find Out?

1 What are three ways that scientists might organize their data?

2 Why do scientists analyze their data?

Evaluate Explanations

Rico observed two water balloons of different sizes. He thought that the bigger water balloon would have more mass because it held more water. He decided to do an investigation to find out.

First Rico used a balance to measure the mass of the small water balloon. Then he measured the mass of the big water balloon. Rico measured each balloon several times to be sure that his measurements were accurate. He organized his data in a table.

Mass of Water Balloons		
Trial	Mass of Large Balloon	Mass of Small Balloon
1	615 g	246 g
2	615 g	246 g
3	615 g	246 g

Analyze Rico's data. Write a paragraph telling whether you agree with Rico's idea that the bigger water balloon has more mass. Use Rico's data to support your argument.

ACKNOWLEDGMENTS Grateful acknowledgment is given to the authors, artists, photographers, museums, publishers, and agents for permission to reprint copyrighted material. Every effort has been made to secure the appropriate permission. If any omissions have been made or if corrections are required, please contact the Publisher.

ILLUSTRATOR CREDITS 29 Susan Anderson. 70 Precision Graphics. 71 Precision Graphics. 73 Paul Dolan. All maps by Mapping Specialists.

PHOTOGRAPHIC CREDITS Front, Back Cover Mark Thiessen/National Geographic Image Collection. **1-2, 3-4, 5-6, 7-8, 9-10** (t) Mark Thiessen/National Geographic Image Collection. **10** Tony Campbell/Shutterstock. **13** (tl) Karen Kuehn/National Geographic Image Collection. (tr) Eremin Sergey/Shutterstock. **14** (bg) Tony Campbell/Shutterstock. (inset, t) Rui Saraiva/Shutterstock. (inset, b) Anton Foltin/Shutterstock. **16** (c) Dmitry Naumov/Shutterstock. (b) Thomas M Perkins/Shutterstock. **16-17** (t) PhotoDisc/Getty Images. **18** (t) PhotoDisc/Getty Images. **19** (b) irishman/Shutterstock. **20** (tbl) János Gehring/Shutterstock. (tbr) Debra McGuire/iStockphoto. (c) Hill Street Studios/Blend Images/Alamy Images. (bl) Socrates/Shutterstock. (blc) bojan fatur/iStockphoto. (br) Dominator/Shutterstock. **20-21, 22** (t) Andrew Williams/Shutterstock. **24** (bl) Ace Stock Limited/Alamy Images. **24-25** (t) Bianca Lavies/National Geographic Image Collection. **26** (t) Bianca Lavies/National Geographic Image Collection. (c) Digital Vision/Getty Images. **27** (b) Frederick R. Matzen/Shutterstock. **28** (t) Ismael Montero Verdu/Shutterstock. (cl) Jonathan Blair/National Geographic Image Collection. (bl) Robert Landau/Alamy Images. **30** (t) Ismael Montero Verdu/Shutterstock. **31** (b) O. Louis Mazzatenta/National Geographic Image Collection. **32, 34** (t) Eric Gevaert/Shutterstock. (bl) itay uri/Shutterstock. **035** (b) Iakov Kalinin/Shutterstock. **36** (t) John Anderson/iStockphoto. **38** (t) John Anderson/iStockphoto. (b) Paul Zahl/National Geographic Image Collection. **39** (b) Tim Laman/National Geographic Image Collection. **40** (l) David Cook / blueshiftstudios/Alamy Images. **40-41** (t) Jhaz Photography/Shutterstock. **42** (inset, r) cameilia/Shutterstock. (t) Jhaz Photography/Shutterstock. **43** (b) Joseph H. Bailey/National Geographic Image Collection. **44** (cl) Chase Swift/iStockphoto. (cr) Michael Stubblefield/iStockphoto. (bl) Jemini Joseph/iStockphoto. (br) Ken Canning/iStockphoto. **44-45** (t) ImageState Royalty Free/Alamy Images. **45** (bg) Radius Images/Alamy Images. (inset) David Young-Wolff/PhotoEdit. **46** Corbis Premium RF/Alamy Images. **46-47** (t) ImageState Royalty Free/Alamy Images. **47** (b) Jeremy Woodhouse/PhotoDisc/Getty Images. **48** (b) Alan D. Carey/PhotoDisc/Getty Images. **48-49, 50** (t) optimarc/Shutterstock. **51** (b) Corel. **52** (t) kwest/Shutterstock. (b) JGI/Blend Images/Getty Images. **54** (t) kwest/Shutterstock. **55** (b) ex0rzist/Shutterstock. **56-57, 58** (t) Anette Linnea Rasmussen/Shutterstock. **59** (b) DigitalStock/Corbis. **60** (b) Nancy Kennedy/Shutterstock. **60-61** (t) Els Jooren/Shutterstock. **61** (b) Chris Curtis/Shutterstock. **62** (c) Anton Foltin/Shutterstock. (b) Corbis. **62-63** (t) Mette Brandt/Shutterstock. **64-69** (t) Mette Brandt/Shutterstock. **70** (t) RandalFung/Corbis RF/Alamy Images. **71** Sheri Lefty/iStockphoto. **72** Ju-Lee/iStockphoto. **73** (t) RandalFung/Corbis RF/Alamy Images. **74** (t) Cartesia/Photodisc/Getty Images. **75** (tl) Shioguchi/Getty Images. (tr) Orange Line Media/Shutterstock. (b) PhotoDisc/Getty Images. **76** (bl) Stockbyte/Getty Images. (bc) Ingram Publishing/Superstock. (br) MetaTools. **79** Randy Santos/Superstock/Photolibrary. **80** (t) Jason Edwards/National Geographic Image Collection. (c) Lawrence Manning/Corbis. (b) pjmorley/Shutterstock. **82** (t) Jason Edwards/National Geographic Image Collection. **83** (b) Tim Fitzharris/Minden Pictures/National Geographic Image Collection. **84** (t) Chernetskiy/Shutterstock. **85** Mark Scheuern/Alamy Images. **87** (t) Chernetskiy/Shutterstock. (b) Ingram Publishing RF/Photolibrary. **88** (t) Radius Images/Photolibrary. (c, b) Frank Zullo/Photo Researchers, Inc.. **90** (t) Radius Images/Photolibrary. **91** (b) Mark Goddard/iStockphoto. **92** (t) BrandX/Jupiterimages. (bl) PhotoDisc/Getty Images. (br) MetaTools. **94** (t) BrandX/Jupiterimages. **95** (b) Ingrid Balabanova/Shutterstock. **96** (t) NASA/GSFC/AIA. **98** (t) NASA/GSFC/AIA. (c) Jani Bryson/iStockphoto. **99** (t) Melba Photo Agency/Alamy Images. **100** (b) Mark Andersen/Getty Images. **100-101, 102** (t) DigitalStock/Corbis. **103** (b) PhotoDisc/Getty Images. **104** (t) NASA/JPL-Caltech/S. Carey (SSC/Caltech)/JPL (NASA). (c) amana images inc./Alamy Images. (bl) Germany Feng/Shutterstock. (bc) Jim Sugar/Corbis. (br) Intraclique LLC/Shutterstock. **106** (t) NASA/JPL-Caltech/S. Carey (SSC/Caltech)/JPL (NASA). **107** (b) Gordon Garradd/Photo Researchers, Inc.. **108** (c) Mike Dunning/Dorling Kindersley Ltd. Picture Library. **108-109, 110** (t) Frans Lanting/National Geographic Image Collection. **111** (b) Frans Lanting/National Geographic Image Collection. **112-113, 114** (t) Anat-oli/Shutterstock. **115** (b) Joyce Marrero/Shutterstock. **116** (bl) Stockbyte/Getty Images. (bcl) Premier Edition Image Library/Superstock. (bcr) Stockbyte/Getty Images. (br) Stockbyte/Getty Images. **116-117** (t) Mariusz S. Jurgielewicz/Shutterstock. **118** (t) Mariusz S. Jurgielewicz/Shutterstock. **119** (b) Tatiana Grozetskaya/Shutterstock. **120, 122** (t) WDG Photo/Shutterstock. **123** (b) Keenpress/National Geographic Image Collection. **124** (t) Jim Richardson/National Geographic Image Collection. (c) PhotoDisc/Getty Images. (b) jeff Metzger/Shutterstock. **125** (tl) David Alan Harvey/National Geographic Image Collection. (tr) Pichugin Dmitry/Shutterstock. (bl) Mick Roessler/Corbis. (br) Hugo Canabi/Iconotec/Alamy Images. **126** (t) Jim Richardson/National

Geographic Image Collection. **127** (b) NASA Human Space Flight Gallery. **128-129, 130** (t) javarman/Shutterstock. **131** (b) Dr Fallow/Shutterstock. **132-133, 134** (t) DigitalStock/Corbis. **135** (b) Nigel Pavitt/Corbis. **136** (b) Medford Taylor/National Geographic Image Collection. **136-137, 138** (t) Aimin Tang/iStockphoto. **139** (b) Danny Zhan/Alamy Images. **140** (b) Duncan Usher/Foto Natura/Minden Pictures/National Geographic Image Collection. **140-141** (t) Elena Elisseeva/Shutterstock. **142** (t) Elena Elisseeva/Shutterstock. **143** (b) PhotoDisc/Getty Images. **144** (t) Geoff du Feu/Alamy Images. (b) Andersen Ross/Blend Images/Getty Images. **146** (t) Geoff du Feu/Alamy Images. **147** (b) PhotoDisc/Getty Images. **148-149** (t) image 100. **149** (b) moodboard/Alamy Images. **150-151** (t) PhotoDisc/Getty Images. **152-157** (t) PhotoDisc/Getty Images. **158** (t) Serg64/Shutterstock. (c) SOHO (ESA & NASA). (b) Brand X/Corbis. **159** Brooks Kraft/Corbis. **160** Tina Stallard/Getty Images. **161** (t) Serg64/Shutterstock. **166-167, 168** (t) SandiMako/Shutterstock. **169** (bl) Ted Kinsman/Photo Researchers, Inc.. (br) Morozova Oxana/Shutterstock. **170-171, 172** (t) Jupiterimages. **173** (b) James Steidl/Shutterstock. **174** (t) Gerry Ellis/Minden Pictures. (b) James P. Blair/National Geographic Image Collection. **176** (t) Gerry Ellis/Minden Pictures. **177** (t) Stockbyte/Getty Images. **178** (t) Ian Shaw/Alamy Images. (c) Julien/Shutterstock. (b) Science Photo Library/Alamy Images. **179** (l) Jack schiffer/Shutterstock. **181** (bg) John Foxx Images/Imagestate. **182-183, 184** (t) Ian Shaw/Alamy Images. **185** (b) bbbb/Shutterstock. **186-187, 188** (t) Vladimir L./Shutterstock. **189** (b) Bartosz Hadyniak/iStockphoto. **190** (t) Nancy Tripp/Shutterstock. **192** (t) Nancy Tripp/Shutterstock. **193** (b) CoolR/Shutterstock. **194** (t) Brent Walker/Shutterstock. (c) Ilin Sergey/Shutterstock. **196** (t) Brent Walker/Shutterstock. **197** (b) Bruce Bean/iStockphoto. **198** (l) ImageState/Alamy Images. **198-199, 200** (t) teacept/Shutterstock. **201** (t) Stanislas Merlin/Jupiterimages. **202** (b) malcolm romain/iStockphoto. **202-203, 204** (t) saied shahin kiya/Shutterstock. **205** (t) Yu Zhang/iStockphoto. **206, 208** (t) nikkytok/Shutterstock. (b) bonnie jacobs/iStockphoto. **209** (b) Tokar Dima/Shutterstock. **210** (t) Aron Hsiao/Shutterstock. (b) hd connelly/Shutterstock. **211** (t) Aron Hsiao/Shutterstock. (b) David Touchtone/Shutterstock. **212** (b) Premier Edition Image Library/Superstock. **212-219** (t) Steve Cole/PhotoDisc/Getty Images. **220** (t) PhotoDisc/Getty Images. (c) Yuri Arcurs/Shutterstock. (b) Digital Vision/Noel Hendrickson/Getty Images. **221** (t) Wouter Tolenaars/Shutterstock. (b) Jon Levy/Getty Images. **222** Laurence Gough/iStockphoto. **223** (t) PhotoDisc/Getty Images.

PROGRAM AUTHORS Judith Sweeney Lederman, Ph.D., Director of Teacher Education and Associate Professor of Science Education, Department of Mathematics and Science Education, Illinois Institute of Technology, Chicago, Illinois; Randy Bell, Ph.D., Associate Professor of Science Education, University of Virginia, Charlottesville, Virginia; Malcolm B. Butler, Ph.D., Associate Professor of Science Education, University of South Florida, St. Petersburg, Florida; Kathy Cabe Trundle, Ph.D., Associate Professor of Early Childhood Science Education, The Ohio State University, Columbus, Ohio; David W. Moore, Ph.D., Professor of Education, College of Teacher Education and Leadership, Arizona State University, Tempe, Arizona

2015 Impression

National Geographic Learning | Cengage Learning
NGL.Cengage.com

888-915-3276

Printed in the USA. RR Donnelley, Menasha, WI

ISBN: 978-1-3051-2043-3

15 16 17 18 19 20 21 22 23

10 9 8 7 6 5 4 3

224